Modular Representation Theory of Finite Groups

Peter Schneider

Modular Representation Theory of Finite Groups

Peter Schneider
Department of Mathematics
University of Münster
Münster
Germany

ISBN 978-1-4471-4831-9 ISBN 978-1-4471-4832-6 (eBook)
DOI 10.1007/978-1-4471-4832-6
Springer Dordrecht Heidelberg New York London

Library of Congress Control Number: 2012954001

Mathematics Subject Classification: 20C20, 20C05

Printed on acid-free paper

Springer is part of Springer Science+Business Media (www.springer.com)

Preface

The nature of the representation theory of a finite group G in (finite-dimensional) vector spaces over some field k depends very much on the relation between the order $|G|$ of the group G and the characteristic $\operatorname{char}(k)$ of the field k. If $\operatorname{char}(k)$ does not divide $|G|$ then all representations are semisimple, i.e. are direct sums of irreducible representations. The reason for this is the semisimplicity of the group algebra $k[G]$ in this situation. By the modular representation theory of G one means, on the other hand, the case where $\operatorname{char}(k)$ is a divisor of $|G|$ (so that, in particular, $\operatorname{char}(k)$ must be a prime number). The group algebra $k[G]$ now may be far from being semisimple. In the extreme case, for example, where $|G|$ is a power of $\operatorname{char}(k)$, it is a local ring; there is then a single irreducible representation, which is the trivial one, whereas the structure of a general representation will still be very complicated. As a consequence a whole range of additional tools have to be developed and used in the course of the investigation. To mention some, there is the systematic use of Grothendieck groups (Chap. 2) as well as Green's direct analysis of indecomposable representations (Chap. 4). There also is the strategy of writing the category of all $k[G]$-modules as the direct product of certain subcategories, the so-called blocks of G, by using the action of the primitive idempotents in the center of $k[G]$. Brauer's approach then establishes correspondences between the blocks of G and blocks of certain subgroups of G (Chap. 5), the philosophy being that one is thereby reduced to a simpler situation. This allows us, in particular, to measure how nonsemisimple a category a block is by the size and structure of its so-called defect group. Beginning in Sect. 4.4 all these concepts are made explicit for the example of the group $G = \mathrm{SL}_2(\mathbb{F}_p)$.

The present book is to be thought of as an introduction to the major tools and strategies of modular representation theory. Its content was taught during a course lasting the full academic year 2010/2011 at Münster. Some basic algebra together with the semisimple case were assumed to be known, although all facts to be used are restated (without proofs) in the text. Otherwise the book is entirely self-contained. The references [1–10] provide a complete list of the sources I have drawn upon. Of course, there already exist several textbooks on the subject. The older ones like [5] and [6] are written in a mostly group theoretic language. The beautiful book [1] develops the theory entirely from the module theoretic point of view but leaves out completely the comparison with group theoretic concepts. For example, the concept of defect groups can be introduced either purely group theoretically or purely module theoretically. To my knowledge all existing books essentially restrict themselves to a discussion of one of these approaches only. Although my presentation is strongly biased towards the module theoretic point of view, I make an attempt to strike a certain balance by also showing the reader the other aspect. In particular, in the case of defect groups a detailed proof of the equivalence of the two approaches will be given.

This book is not addressed to experts. It does not discuss any very advanced aspects nor any specialized results of the theory. The aim is to familiarize students at the masters level with the basic results, tools, and techniques of a beautiful and important algebraic theory, hopefully enabling them to subsequently pursue their own more specialized problems.

I wish to thank T. Schmidt for carefully reading a first draft and I. Reckermann and G. Dierkes for their excellent typesetting of the manuscript.

Münster, Germany

Peter Schneider

Contents

Chapter 1
Prerequisites in Module Theory

Let R be an arbitrary (not necessarily commutative) ring (with unit). By an R-module we will always mean a left R-module. All ring homomorphisms respect the unit element, but a subring may have a different unit element.

1.1 Chain Conditions and More

For an R-module M we have the notions of being

finitely generated, artinian, noetherian, simple, and semisimple.

The ring R is called left artinian, resp. left noetherian, resp. semisimple, if it has this property as a left module over itself.

Proposition 1.1.1

i. *The R-module M is noetherian if and only if any submodule of M is finitely generated.*
ii. *Let $L \subseteq M$ be a submodule; then M is artinian, resp. noetherian, if and only if L and M/L are artinian, resp. noetherian.*
iii. *If R is left artinian, resp. left noetherian, then every finitely generated R-module M is artinian, resp. noetherian.*
iv. *If R is left noetherian then an R-module M is noetherian if and only if it is finitely generated.*

Proposition 1.1.2 (Jordan–Hölder) *For any R-module M the following conditions are equivalent:*

i. *M is artinian and noetherian;*
ii. *M has a composition series $\{0\} = M_0 \subseteq M_1 \subseteq \cdots \subseteq M_n = M$ such that all M_i/M_{i-1} are simple R-modules.*

P. Schneider, *Modular Representation Theory of Finite Groups*,
DOI 10.1007/978-1-4471-4832-6_1, © Springer-Verlag London 2013

In this case two composition series $\{0\} = M_0 \subseteq M_1 \subseteq \cdots \subseteq M_n = M$ *and* $\{0\} = L_0 \subseteq L_1 \subseteq \cdots \subseteq L_m = M$ *satisfy* $n = m$ *and* $L_i/L_{i-1} \cong M_{\sigma(i)}/M_{\sigma(i)-1}$, *for any* $1 \leq i \leq m$, *where* σ *is an appropriate permutation of* $\{1, \ldots, n\}$.

An R-module M which satisfies the conditions of Proposition 1.1.2 is called *of finite length* and the integer $l(M) := n$ is its *length*. Let

$$\hat{R} := \text{set of all isomorphism classes of simple } R\text{-modules.}$$

For $\tau \in \hat{R}$ and an R-module M the τ-isotypic component of M is

$$M(\tau) := \text{sum of all simple submodules of } M \text{ in } \tau.$$

Lemma 1.1.3 *For any* R*-module homomorphism* $f : L \longrightarrow M$ *we have* $f(L(\tau)) \subseteq M(\tau)$.

Proposition 1.1.4

i. *For any* R*-module* M *the following conditions are equivalent*:
   a. M *is semisimple, i.e. isomorphic to a direct sum of simple* R*-modules*;
   b. M *is the sum of its simple submodules*;
   c. *every submodule of* M *has a complement.*
ii. *Submodules and factor modules of semisimple modules are semisimple.*
iii. *If* R *is semisimple then any* R*-module is semisimple.*
iv. *Any* τ*-isotypic component* $M(\tau)$ *of any* R*-module* M *is semisimple.*
v. *If the* R*-module* M *is semisimple then* $M = \bigoplus_{\tau \in \hat{R}} M(\tau)$.
vi. *If* R *is semisimple then* $R = \prod_{\tau \in \hat{R}} R(\tau)$ *as rings.*

Lemma 1.1.5 *Any* R*-module* M *contains a unique maximal submodule which is semisimple (and which is called the socle* $\mathrm{soc}(M)$ *of* M*).*

Proof We define $\mathrm{soc}(M) := \sum_{\tau \in \hat{R}} M(\tau)$. By Proposition 1.1.4.i the submodule $\mathrm{soc}(M)$ is semisimple. On the other hand if $L \subseteq M$ is any semisimple submodule then $L = \sum_\tau L(\tau)$ by Proposition 1.1.4.v. But $L(\tau) \subseteq M(\tau)$ by Lemma 1.1.3 and hence $L \subseteq \mathrm{soc}(M)$. □

Definition An R-module M is called decomposable if there exist nonzero submodules $M_1, M_2 \subseteq M$ such that $M = M_1 \oplus M_2$; correspondingly, M is called indecomposable if it is nonzero and not decomposable.

Lemma 1.1.6 *If* M *is artinian or noetherian then* M *is the direct sum of finitely many indecomposable submodules.*

Proof We may assume that $M \neq \{0\}$. *Step 1:* We claim that M has a nonzero indecomposable direct summand N. If M is artinian take a minimal element N of the set of all nonzero direct summands of M. If M is noetherian let N' be a maximal element of the set of all direct summands $\neq M$ of M, and take N such that $M = N' \oplus N$. *Step 2:* By Step 1 we find $M = M_1 \oplus M_1'$ with $M_1 \neq \{0\}$ indecomposable. If $M_1' \neq \{0\}$ then similarly $M_1' = M_2 \oplus M_2'$ with $M_2 \neq \{0\}$ indecomposable. Inductively we obtain in this way a strictly increasing sequence

$$\{0\} \subsetneqq M_1 \subsetneqq M_1 \oplus M_2 \subsetneqq \cdots$$

as well as a strictly decreasing sequence

$$M_1' \supsetneqq M_2' \supsetneqq \cdots$$

of submodules of M. Since one of the two (depending on M being artinian or noetherian) stops after finitely many steps we must have $M_n' = \{0\}$ for some n. Then $M = M_1 \oplus \cdots \oplus M_n$ is the direct sum of the finitely many indecomposable submodules $M_1, \ldots, M_n$. □

Exercise The $\mathbb{Z}$-modules $\mathbb{Z}$ and $\mathbb{Z}/p^n\mathbb{Z}$, for any prime number p and any $n \geq 1$, are indecomposable.

1.2 Radicals

The radical of an R-module M is the submodule

$$\mathrm{rad}(M) := \text{intersection of all maximal submodules of } M.$$

The Jacobson radical of R is $\mathrm{Jac}(R) := \mathrm{rad}(R)$.

Proposition 1.2.1

i. *If $M \neq \{0\}$ is finitely generated then* $\mathrm{rad}(M) \neq M$.
ii. *If M is semisimple then* $\mathrm{rad}(M) = \{0\}$.
iii. *If M is artinian with* $\mathrm{rad}(M) = \{0\}$ *then M is semisimple.*
iv. *The Jacobson radical*

$$\begin{aligned} \mathrm{Jac}(R) &= \textit{intersection of all maximal left ideals of } R \\ &= \textit{intersection of all maximal right ideals of } R \\ &= \{a \in R : 1 + Ra \subseteq R^\times\} \end{aligned}$$

is a two-sided ideal of R.
v. *Any left ideal which consists of nilpotent elements is contained in* $\mathrm{Jac}(R)$.
vi. *If R is left artinian then the ideal* $\mathrm{Jac}(R)$ *is nilpotent and the factor ring* $R/\mathrm{Jac}(R)$ *is semisimple.*
vii. *If R is left artinian then* $\mathrm{rad}(M) = \mathrm{Jac}(R)M$ *for any finitely generated R-module M.*

Corollary 1.2.2 *If R is left artinian then any artinian R-module is noetherian and, in particular, R is left noetherian and any finitely generated R-module is of finite length.*

Lemma 1.2.3 (Nakayama) *If $L \subseteq M$ is a submodule of an R-module M such that M/L is finitely generated, then $L + \mathrm{Jac}(R)M = M$ implies that $L = M$.*

Lemma 1.2.4 *If R is left noetherian and $R/\mathrm{Jac}(R)$ is left artinian then $R/\mathrm{Jac}(R)^m$ is left noetherian and left artinian for any $m \geq 1$.*

Proof It suffices to show that $R/\mathrm{Jac}(R)^m$ is noetherian and artinian as an R-module. By Proposition 1.1.1.ii it further suffices to prove that the R-module $\mathrm{Jac}(R)^m/\mathrm{Jac}(R)^{m+1}$, for any $m \geq 0$, is artinian. Since R is left noetherian it certainly is a finitely generated R-module and hence a finitely generated $R/\mathrm{Jac}(R)$-module. The claim therefore follows from Proposition 1.1.1.iii. □

Lemma 1.2.5 *The Jacobson ideal of the matrix ring $M_{n\times n}(R)$, for any $n \in \mathbb{N}$, is the ideal $\mathrm{Jac}(M_{n\times n}(R)) = M_{n\times n}(\mathrm{Jac}(R))$ of all matrices with entries in $\mathrm{Jac}(R)$.*

1.3 I-Adic Completeness

We begin by introducing the following general construction. Let

$$(M_{n+1} \xrightarrow{f_n} M_n)_{n\in\mathbb{N}}$$

be a sequence of R-module homomorphisms, usually visualized by the diagram

$$\cdots \xrightarrow{f_{n+1}} M_{n+1} \xrightarrow{f_n} M_n \xrightarrow{f_{n-1}} \cdots \xrightarrow{f_1} M_1.$$

In the direct product module $\prod_{n\in\mathbb{N}} M_n$ we then have the submodule

$$\varprojlim M_n := \left\{ (x_n)_n \in \prod_n M_n : f_n(x_{n+1}) = x_n \text{ for any } n \in \mathbb{N} \right\},$$

which is called the *projective limit* of the above sequence. Although suppressed in the notation the construction depends crucially, of course, on the homomorphisms f_n and not only on the modules M_n.

Let us fix now a two-sided ideal $I \subseteq R$. Its powers form a descending sequence

$$R \supseteq I \supseteq I^2 \supseteq \cdots \supseteq I^n \supseteq \cdots$$

of two-sided ideals. More generally, for any R-module M we have the descending sequence of submodules

$$M \supseteq IM \supseteq I^2M \supseteq \cdots \supseteq I^nM \supseteq \cdots$$

and correspondingly the sequence of residue class projections

$$\cdots \longrightarrow M/I^{n+1}M \xrightarrow{\text{pr}} M/I^nM \longrightarrow \cdots \longrightarrow M/I^2M \xrightarrow{\text{pr}} M/IM.$$

We form the projective limit of the latter

$$\varprojlim M/I^nM = \left\{ \left(x_n + I^nM\right)_n \in \prod_n M/I^nM : x_{n+1} - x_n \in I^nM \text{ for any } n \in \mathbb{N} \right\}.$$

There is the obvious R-module homomorphism

$$\begin{aligned} \pi_M^I: \quad M &\longrightarrow \varprojlim M/I^nM \\ x &\longmapsto \left(x + I^nM\right)_n. \end{aligned}$$

Definition The R-module M is called I-adically separated, resp. I-adically complete, if π_M^I is injective (i.e. if $\bigcap_n I^nM = \{0\}$), resp. is bijective.

Exercise If $I^{n_0}M = \{0\}$ for some $n_0 \in \mathbb{N}$ then M is I-adically complete.

Lemma 1.3.1 *Let $J \subseteq R$ be another two-sided ideal such that $I^{n_0} \subseteq J \subseteq I$ for some $n_0 \in \mathbb{N}$; then M is J-adically separated, resp. complete, if and only if it is I-adically separated, resp. complete.*

Proof We have

$$I^{nn_0}M \subseteq J^nM \subseteq I^nM \quad \text{for any } n \in \mathbb{N}.$$

This makes obvious the separatedness part of the assertion. Furthermore, by the right-hand inclusions we have the well-defined map

$$\begin{aligned} \alpha: \quad \varprojlim M/J^nM &\longrightarrow \varprojlim M/I^nM \\ \left(x_n + J^nM\right)_n &\longmapsto \left(x_n + I^nM\right)_n. \end{aligned}$$

Since $\alpha \circ \pi_M^J = \pi_M^I$ it suffices for the completeness part of the assertion to show that α is bijective. That $\alpha(\xi) = 0$ for $(x_n + J^nM)_n \in \varprojlim M/J^nM$ means that $x_n \in I^nM$ for any $n \in \mathbb{N}$. Hence $x_{nn_0} \in I^{nn_0}M \subseteq J^nM$. On the other hand $x_{nn_0} - x_n \in J^nM$. It follows that $x_n \in J^nM$ for any n, i.e. that $\xi = 0$. For the surjectivity of α let $(y_n + I^nM)_n \in \varprojlim M/I^nM$ be any element. We define $x_n := y_{nn_0}$. Then $x_{n+1} - x_n = y_{nn_0+n_0} - y_{nn_0} \in I^{nn_0}M \subseteq J^nM$ and hence $(x_n + J^nM)_n \in \varprojlim M/J^nM$. Moreover, $x_n + I^nM = y_{nn_0} + I^nM = y_n + I^nM$ which shows that $\alpha((x_n + J^nM)_n) = (y_n + I^nM)_n$. □

Lemma 1.3.2 *If R is I-adically complete then $I \subseteq \mathrm{Jac}(R)$.*

Proof Let $a \in I$ be any element. Then

$$\left(1 + a + \cdots + a^{n-1} + I^n\right)_n \in \varprojlim R/I^n.$$

Therefore, by assumption, there is an element $c \in R$ such that

$$c + I^n = 1 + a + \cdots + a^{n-1} + I^n \quad \text{for any } n \in \mathbb{N}.$$

It follows that

$$(1-a)c \in \bigcap_n (1-a)\left(1 + a + \cdots + a^{n-1}\right) + I^n = \bigcap_n 1 - a^n + I^n = \{1\}$$

and similarly that $c(1-a) = 1$. This proves that $1 + I \subseteq R^\times$. The assertion now follows from Proposition 1.2.1.iv. □

Lemma 1.3.3 *Let $f : M \longrightarrow N$ be a surjective R-module homomorphism and suppose that M is I-adically complete; if N is I-adically separated then it already is I-adically complete.*

Proof The R-module homomorphism

$$\begin{aligned} \phi: \quad \varprojlim M/I^n M &\longrightarrow \varprojlim N/I^n N \\ \left(x_n + I^n M\right)_n &\longmapsto \left(f(x_n) + I^n N\right)_n \end{aligned}$$

fits into the commutative diagram

$$\begin{array}{ccc} M & \xrightarrow{\quad f \quad} & N \\ {\scriptstyle \alpha_M^I} \downarrow \cong & & \downarrow {\scriptstyle \alpha_N^I} \\ \varprojlim M/I^n M & \xrightarrow{\quad \phi \quad} & \varprojlim N/I^n N. \end{array}$$

Hence it suffices to show that ϕ is surjective. Let $(y_n + I^n N)_n \in \varprojlim N/I^n N$ be an arbitrary element. By the surjectivity of f we find an $x_1 \in M$ such that $f(x_1) = y_1$. We now proceed by induction and assume that elements $x_1, \ldots, x_n \in M$ have been found such that

$$x_{j+1} - x_j \in I^j M \quad \text{for } 1 \le j < n$$

and

$$f(x_j) + I^j N = y_j + I^j N \quad \text{for } 1 \le j \le n.$$

We choose any $x'_{n+1} \in M$ such that $f(x'_{n+1}) = y_{n+1}$. Then

$$f\left(x'_{n+1} - x_n\right) \in y_{n+1} - y_n + I^n N = I^n N.$$

Since the restricted map $I^nM \xrightarrow{f} I^nN$ still is surjective we find an element $z \in \ker(f)$ such that

$$x'_{n+1} - x_n - z \in I^nM.$$

Therefore, if we put $x_{n+1} := x'_{n+1} - z$ then $x_{n+1} - x_n \in I^nM$ and $f(x_{n+1}) = f(x'_{n+1}) = y_{n+1}$. □

Proposition 1.3.4 *Suppose that R is commutative and noetherian and that $I \subseteq \mathrm{Jac}(R)$; then any finitely generated R-module M is I-adically separated.*

Proof We have to show that the submodule $M_0 := \bigcap_n I^nM$ is zero. Since R is noetherian M_0 is finitely generated by Proposition 1.1.1. We therefore may apply the Nakayama lemma 1.2.3 to M_0 (and its submodule $\{0\}$) and see that it suffices to show that $IM_0 = M_0$. Obviously $IM_0 \subseteq M_0$. Again since R is noetherian we find a submodule $IM_0 \subseteq L \subseteq M$ which is maximal with respect to the property that $L \cap M_0 = IM_0$.

In an intermediate step we establish that for any $a \in I$ there is an integer $n_0(a) \geq 1$ such that

$$a^{n_0(a)}M \subseteq L.$$

Fixing a we put $M_j := \{x \in M : a^jx \in L\}$ for $j \geq 1$. Since R is commutative the $M_j \subseteq M_{j+1}$ form an increasing sequence of submodules of M. Since R is noetherian this sequence

$$M_1 \subseteq \cdots \subseteq M_{n_0(a)} = M_{n_0(a)+1} = \cdots$$

has to stabilize. We trivially have $IM_0 \subseteq (a^{n_0(a)}M + L) \cap M_0$. Consider any element $x_0 = a^{n_0(a)}x + y$, with $x \in M$ and $y \in L$, in the right-hand side. We have $ax_0 \in IM_0 \subseteq L$ and $ay \in L$, hence $a^{n_0(a)+1}x = ax_0 - ay \in L$ or equivalently $x \in M_{n_0(a)+1}$. But $M_{n_0(a)+1} = M_{n_0(a)}$ so that $a^{n_0(a)}x \in L$ and consequently $x_0 \in L \cap M_0 = IM_0$. This shows that $IM_0 = (a^{n_0(a)}M + L) \cap M_0$ holds true. The maximality of L then implies that $a^{n_0(a)}M \subseteq L$.

The ideal I in the noetherian ring R can be generated by finitely many elements $a_1, \ldots, a_r$. We put $n_0 := \max_i n_0(a_i)$ and $n_1 := rn_0$. Then

$$I^{n_1} = (Ra_1 + \cdots + Ra_r)^{rn_0} \subseteq Ra_1^{n_0} + \cdots + Ra_r^{n_0} \subseteq Ra_r^{n_0(a_1)} + \cdots + Ra_r^{n_0(a_r)}$$

and hence $I^{n_1}M \subseteq L$ which implies

$$M_0 = \bigcap_n I^nM \subseteq I^{n_1}M \subseteq L \quad \text{and therefore} \quad M_0 = L \cap M_0 = IM_0.$$

□

Let R_0 be a commutative ring. If $\alpha : R_0 \longrightarrow Z(R)$ is a ring homomorphism into the center $Z(R)$ of R then we call R an R_0-algebra (with respect to α). In particular, R then is an R_0-module. More generally, by restriction of scalars, any R-module also is an R_0-module.

For any R-module M we have the two endomorphism rings $\mathrm{End}_R(M) \subseteq \mathrm{End}_{R_0}(M)$. Both are R_0-algebras with respect to the homomorphism

$$\begin{aligned} R_0 &\longrightarrow \mathrm{End}_R(M) \\ a_0 &\longmapsto a_0 \cdot \mathrm{id}_M . \end{aligned}$$

Lemma 1.3.5 *Suppose that R is an R_0-algebra which as an R_0-module is finitely generated, and assume R_0 to be noetherian; we than have:*

i. *R is left and right noetherian;*
ii. *for any finitely generated R-module M its ring $\mathrm{End}_R(M)$ of endomorphisms is left and right noetherian and is finitely generated as an R_0-module;*
iii. $\mathrm{Jac}(R_0)R \subseteq \mathrm{Jac}(R)$.

Proof i. Any left or right ideal of R is a submodule of the noetherian R_0-module R and hence is finitely generated (cf. Proposition 1.1.1).

ii. *Step 1:* We claim that, for any finitely generated R_0-module M_0, the R_0-module $\mathrm{End}_{R_0}(M_0)$ is finitely generated. Let $x_1, \ldots, x_r$ be generators of the R_0-module M_0. Then

$$\begin{aligned} \mathrm{End}_{R_0}(M_0) &\longrightarrow M_0 \oplus \cdots \oplus M_0 \\ f &\longmapsto \big(f(x_1), \ldots, f(x_r)\big) \end{aligned}$$

is an injective R_0-module homomorphism. The right-hand side is finitely generated by assumption and so is then the left-hand side since R_0 is noetherian. *Step 2:* By assumption M is finitely generated over R, and R is finitely generated over R_0. Hence M is finitely generated over R_0, Step 1 applies, and $\mathrm{End}_{R_0}(M)$ is a finitely generated R_0-module. Since R_0 is noetherian the submodule $\mathrm{End}_R(M)$ is finitely generated as well. Furthermore, since any left or right ideal of $\mathrm{End}_R(M)$ is an R_0-submodule the ring $\mathrm{End}_R(M)$ is left and right noetherian.

iii. Set $L \subseteq R$ be a maximal left ideal. Then $M := R/L$ is a simple R-module. By ii. the R_0-module $\mathrm{End}_R(M)$ is finitely generated. It is nonzero since M is nonzero. The Nakayama lemma 1.2.3 therefore implies that $\mathrm{Jac}(R_0)\,\mathrm{End}_R(M) \neq \mathrm{End}_R(M)$. But $\mathrm{End}_R(M)$, by Schur's lemma, is a skew field. It follows that $\mathrm{Jac}(R_0)\,\mathrm{End}_R(M) = \{0\}$. Hence

$$\mathrm{Jac}(R_0)M = \mathrm{id}_M\big(\mathrm{Jac}(R_0)M\big) = \mathrm{Jac}(R_0)\,\mathrm{id}_M(M) = \{0\}$$

or equivalently $\mathrm{Jac}(R_0)R \subseteq L$. Since L was arbitrary we obtain the assertion. □

For simplicity we call an R-module complete if it is $\mathrm{Jac}(R)$-adically complete.

Proposition 1.3.6 *Suppose that R is an R_0-algebra which as an R_0-module is finitely generated, and assume that R_0 is noetherian and complete and that $R_0/\mathrm{Jac}(R_0)$ is artinian; for any of the rings $S = R$ or $S = \mathrm{End}_R(M)$, where M is a finitely generated R-module, we then have:*

i. *S is left and right noetherian;*
ii. *$S/\operatorname{Jac}(S)$ is left and right artinian;*
iii. *any finitely generated S-module is complete.*

Proof By Lemma 1.3.5.ii the case $S = R$ contains the case $S = \operatorname{End}_R(M)$. Lemma 1.3.5.i says that R is left and right noetherian. Since R_0 maps to the center of R the right ideal $\operatorname{Jac}(R_0)R$ in fact is two-sided. Being finitely generated as an $R_0/\operatorname{Jac}(R_0)$-module the ring $R/\operatorname{Jac}(R_0)R$ is left and right artinian by Proposition 1.1.1.iii. According to Lemma 1.3.5.iii the ring $R/\operatorname{Jac}(R)$ is a factor ring of $R/\operatorname{Jac}(R_0)R$ and therefore, by Proposition 1.1.1.ii, is left and right artinian as well. Using Proposition 1.2.1.vi it also follows that

$$\operatorname{Jac}(R)^{n_0} \subseteq \operatorname{Jac}(R_0)R \subseteq \operatorname{Jac}(R)$$

for some $n_0 \in \mathbb{N}$. Because of Lemma 1.3.1 it therefore remains to show that any finitely generated R-module N is $\operatorname{Jac}(R_0)R$-adically complete. Since

$$(\operatorname{Jac}(R_0)R)^n N = \operatorname{Jac}(R_0)^n N \quad \text{for any } n \geq 1$$

this further reduces to the statement that any finitely generated R_0-module N is complete. We know from Proposition 1.3.4 that the R_0-module N is $\operatorname{Jac}(R_0)$-adically separated. On the other hand, by finite generation we find, for some $m \in \mathbb{N}$, a surjective R_0-module homomorphism $R_0^m \twoheadrightarrow N$. With R_0 also R_0^m is complete by assumption. We therefore may apply Lemma 1.3.3 and obtain that N is complete. □

1.4 Unique Decomposition

We first introduce the following concept.

Definition 1 A ring A is called local if $A \setminus A^\times$ is a two-sided ideal of A.

We note that a local ring A is nonzero since $1 \in A^\times$ whereas 0 must lie in the ideal $A \setminus A^\times$.

Proposition 1.4.1 *For any nonzero ring A the following conditions are equivalent:*

i. *A is local;*
ii. *$A \setminus A^\times$ is additively closed;*
iii. *$A \setminus \operatorname{Jac}(A) \subseteq A^\times$;*
iv. *$A/\operatorname{Jac}(A)$ is a skew field;*
v. *A contains a unique maximal left ideal;*
vi. *A contains a unique maximal right ideal.*

Proof We note that $A \neq \{0\}$ implies that $\operatorname{Jac}(A) \neq A$ is a proper ideal.

i. $\Longrightarrow$ ii. This is obvious. ii. $\Longrightarrow$ iii. Let $b \in A \setminus A^\times$ be any element. Then also $-b \notin A^\times$. Suppose that $1 + ab \notin A^\times$ for some $a \in A$. Using that $A \setminus A^\times$ is additively closed we first obtain that $-ab \in A^\times$ and in particular that $-a, a \notin A^\times$. It then follows that $a + b \notin A^\times$ and that $1 + a, 1 + b \in A^\times$. But in the identity

$$(1 + ab) + (a + b) = (1 + a)(1 + b)$$

now the right-hand side is contained in $A^\times$ whereas the left-hand side is not. This contradiction proves that $1 + Ab \subseteq A^\times$ and therefore, by Proposition 1.2.1.iv, that $b \in \operatorname{Jac}(A)$. We conclude that $A \subseteq A^\times \cup \operatorname{Jac}(A)$.

iii. $\Longrightarrow$ iv. It immediately follows from iii. that any nonzero element in $A/\operatorname{Jac}(A)$ is a unit.

iv. $\Longrightarrow$ v., vi. Let $L \subseteq A$ be any maximal left, resp. right, ideal. Then $\operatorname{Jac}(A) \subseteq L \subsetneqq A$ by Proposition 1.2.1.iv. Since the only proper left, resp. right, ideal of a skew field is the zero ideal we obtain $L = \operatorname{Jac}(A)$.

v., vi. $\Longrightarrow$ i. The unique maximal left (right) ideal, by Proposition 1.2.1.iv, is necessarily equal to $\operatorname{Jac}(A)$. Let $b \in A \setminus \operatorname{Jac}(A)$ be any element. Then Ab (bA) is not contained in any maximal left (right) ideal and hence $Ab = A$ ($bA = A$). Let $a \in A$ such that $ab = 1$ ($ba = 1$). Since $a \notin \operatorname{Jac}(A)$ we may repeat this reasoning and find an element $c \in A$ such that $ca = 1$ ($ac = 1$). But $c = c(ab) = (ca)b = b$ ($c = (ba)c = b(ac) = b$). It follows that $a \in A^\times$ and then also that $b \in A^\times$. This shows that $A = \operatorname{Jac}(A) \cup A^\times$. But the union is disjoint. We finally obtain that $A \setminus A^\times = \operatorname{Jac}(A)$ is a two-sided ideal. □

We see that in a local ring A the Jacobson radical $\operatorname{Jac}(A)$ is the unique maximal left (right, two-sided) ideal and that $A \setminus \operatorname{Jac}(A) = A^\times$.

Lemma 1.4.2 *If A is nonzero and any element in $A \setminus A^\times$ is nilpotent then the ring A is local.*

Proof Let $b \in A \setminus A^\times$ and let $n \geq 1$ be minimal such that $b^n = 0$. For any $a \in A$ we then have $(ab)b^{n-1} = ab^n = 0$. Since $b^{n-1} \neq 0$ the element ab cannot be a unit in A. It follows that $Ab \subseteq A \setminus A^\times$ and hence, by Proposition 1.2.1.v, that $Ab \subseteq \operatorname{Jac}(A)$. We thus have shown that $A \setminus A^\times \subseteq \operatorname{Jac}(A)$ which, by Proposition 1.4.1.iii, implies that A is local. □

Lemma 1.4.3 (Fitting) *For any R-module M and any $f \in \operatorname{End}_R(M)$ we have:*

i. *If M is noetherian and f is injective then f is bijective;*
ii. *if M is noetherian and f is surjective then f is bijective;*
iii. *if M is of finite length then there exists an integer $n \geq 1$ such that*
   a. $\ker(f^n) = \ker(f^{n+j})$ *for any* $j \geq 0$,
   b. $\operatorname{im}(f^n) = \operatorname{im}(f^{n+j})$ *for any* $j \geq 0$,
   c. $M = \ker(f^n) \oplus \operatorname{im}(f^n)$, *and*
   d. *the induced maps* $\operatorname{id}_M - f : \ker(f^n) \xrightarrow{\cong} \ker(f^n)$ *and* $f : \operatorname{im}(f^n) \xrightarrow{\cong} \operatorname{im}(f^n)$ *are bijective.*

Proof We have the increasing sequence of submodules

$$\ker(f) \subseteq \ker(f^2) \subseteq \cdots \subseteq \ker(f^n) \subseteq \cdots$$

as well as the decreasing sequence of submodules

$$\operatorname{im}(f) \supseteq \operatorname{im}(f^2) \supseteq \cdots \supseteq \operatorname{im}(f^n) \supseteq \cdots.$$

If M is artinian there must exist an $n \geq 1$ such that

$$\operatorname{im}(f^n) = \operatorname{im}(f^{n+1}) = \cdots = \operatorname{im}(f^{n+j}) = \cdots.$$

For any $x \in M$ we then find a $y \in M$ such that $f^n(x) = f^{n+1}(y)$. Hence $f^n(x - f(y)) = 0$. If f is injective it follows that $x = f(y)$. This proves $M = f(M)$ under the assumptions in i.

If M is noetherian there exists an $n \geq 1$ such that

$$\ker(f^n) = \ker(f^{n+1}) = \cdots = \ker(f^{n+j}) = \cdots.$$

Let $x \in \ker(f)$. If f, and hence f^n, is surjective we find a $y \in M$ such that $x = f^n(y)$. Then $f^{n+1}(y) = f(x) = 0$, i.e. $y \in \ker(f^{n+1}) = \ker(f^n)$. It follows that $x = f^n(y) = 0$. This proves $\ker(f) = \{0\}$ under the assumptions in ii.

Assuming that M is of finite length, i.e. artinian and noetherian, we at least know the existence of an $n \geq 1$ satisfying a. and b. To establish c. (for any such n) we first consider any $x \in \ker(f^n) \cap \operatorname{im}(f^n)$. Then $f^n(x) = 0$ and $x = f^n(y)$ for some $y \in M$. Hence $y \in \ker(f^{2n}) = \ker(f^n)$ which implies $x = 0$. This shows that

$$\ker(f^n) \cap \operatorname{im}(f^n) = \{0\}.$$

Secondly let $x \in M$ be arbitrary. Then $f^n(x) \in \operatorname{im}(f^n) = \operatorname{im}(f^{2n})$, i.e. $f^n(x) = f^{2n}(y)$ for some $y \in M$. We obtain $f^n(x - f^n(y)) = f^n(x) - f^{2n}(y) = 0$. Hence

$$x = (x - f^n(y)) + f^n(y) \in \ker(f^n) + \operatorname{im}(f^n).$$

For d. we note that $\mathrm{id}_M - f : \ker(f^n) \longrightarrow \ker(f^n)$ has the inverse $\mathrm{id}_m + f + \cdots + f^{n-1}$. Since $\ker(f) \cap \operatorname{im}(f^n) = \{0\}$ by c., the restriction $f | \operatorname{im}(f^n)$ is injective. Let $x \in \operatorname{im}(f^n)$. Because of b we find a $y \in M$ such that $x = f^{n+1}(y) = f(f^n(y)) \in f(\operatorname{im}(f^n))$. □

Proposition 1.4.4 *For any indecomposable R-module M of finite length the ring* $\mathrm{End}_R(M)$ *is local.*

Proof Since $M \neq \{0\}$ we have $\mathrm{id}_M \neq 0$ and hence $\mathrm{End}_R(M) \neq \{0\}$. Let $f \in \mathrm{End}_R(M)$ be any element which is not a unit, i.e. is not bijective. According to Lemma 1.4.3.iii we have

$$M = \ker(f^n) \oplus \operatorname{im}(f^n) \quad \text{for some } n \geq 1.$$

But M is indecomposable. Hence $\ker(f^n) = M$ or $\operatorname{im}(f^n) = M$. In the latter case f would be bijective by Lemma 1.4.3.iii.d which leads to a contradiction. It follows that $f^n = 0$. This shows that the assumption in Lemma 1.4.2 is satisfied so that $\operatorname{End}_R(M)$ is local. □

Proposition 1.4.5 *Suppose that R is left noetherian, $R/\operatorname{Jac}(R)$ is left artinian, and any finitely generated R-module is complete; then $\operatorname{End}_R(M)$, for any finitely generated indecomposable R-module M, is a local ring.*

Proof We abbreviate $J := \operatorname{Jac}(R)$. By assumption the map

$$\pi_M^J\colon\ M \xrightarrow{\cong} \varprojlim M/J^m M$$

is an isomorphism. Each $M/J^m M$ is a finitely generated module over the factor ring R/J^m which, by Lemma 1.2.4, is left noetherian and left artinian. Hence $M/J^m M$ is a module of finite length by Proposition 1.1.1.iii.

On the other hand for any $f \in \operatorname{End}_R(M)$ the R-module homomorphisms

$$\begin{aligned} f_m\colon\ M/J^m M &\longrightarrow M/J^m M \\ x + J^m M &\longmapsto f(x) + J^m M \end{aligned}$$

are well defined. The diagrams

$$\begin{array}{ccc} M/J^{m+1}M & \xrightarrow{f_{m+1}} & M/J^{m+1}M \\ {\scriptstyle \mathrm{pr}}\downarrow & & \downarrow{\scriptstyle \mathrm{pr}} \\ M/J^m M & \xrightarrow{f_m} & M/J^m M \end{array} \tag{1.4.1}$$

obviously are commutative so that in the projective limit we obtain the R-module homomorphism

$$\begin{aligned} f_\infty\colon\ \varprojlim M/J^m M &\longrightarrow \varprojlim M/J^m M \\ \left(x_m + J^m M\right)_m &\longmapsto \left(f(x_m) + J^m M\right)_m. \end{aligned}$$

Clearly the diagram

$$\begin{array}{ccc} M & \xrightarrow{f} & M \\ {\scriptstyle \pi_M^J}\downarrow{\scriptstyle \cong} & & {\scriptstyle \cong}\downarrow{\scriptstyle \pi_M^J} \\ \varprojlim M/J^m M & \xrightarrow{f_\infty} & \varprojlim M/J^m M \end{array}$$

is commutative. If follows, for example, that f is bijective if and only if f_∞ is bijective.

We now apply Fitting's lemma 1.4.3.iii to each module $M/J^m M$ and obtain an increasing sequence of integers $1 \le n(1) \le \cdots \le n(m) \le \cdots$ such that the triple $(M/J^m M, f_m, n(m))$ satisfies the conditions a.–d. in that lemma. In particular, we have

$$M/J^m M = \ker(f_m^{n(m)}) \oplus \operatorname{im}(f_m^{n(m)}) \quad \text{for any } m \ge 1.$$

The commutativity of (1.4.1) easily implies that we have the sequences of surjective (!) R-module homomorphisms

$$\cdots \xrightarrow{\text{pr}} \ker(f_{m+1}^{n(m+1)}) \xrightarrow{\text{pr}} \ker(f_m^{n(m)}) \xrightarrow{\text{pr}} \cdots \xrightarrow{\text{pr}} \ker(f_1^{n(1)})$$

and

$$\cdots \xrightarrow{\text{pr}} \operatorname{im}(f_{m+1}^{n(m+1)}) \xrightarrow{\text{pr}} \operatorname{im}(f_m^{n(m)}) \xrightarrow{\text{pr}} \cdots \xrightarrow{\text{pr}} \operatorname{im}(f_1^{n(1)}).$$

Be defining

$$X := \varprojlim \ker(f_m^{n(m)}) \quad \text{and} \quad Y := \varprojlim \operatorname{im}(f_m^{n(m)})$$

we obtain the decomposition into R-submodules

$$M \cong \varprojlim M/J^m M = X \oplus Y.$$

But M is indecomposable. Hence $X = \{0\}$ or $Y = \{0\}$. Suppose first that $X = \{0\}$. By the surjectivity of the maps in the corresponding sequence this implies $\ker(f_m^{n(m)}) = \{0\}$ for any $m \ge 1$. The condition d then says that all the f_m, hence f_∞, and therefore f are bijective. If, on the other hand, $Y = \{0\}$ then analogously $\operatorname{im}(f_m^{n(m)}) = \{0\}$ for all $m \ge 1$, and conditions d. implies that $\mathrm{id}_M - f$ is bijective.

So far we have shown that for any $f \in \operatorname{End}_R(M)$ either f or $\mathrm{id}_M - f$ is a unit. To prove our assertion it suffices, by Proposition 1.4.1.ii, to verify that the nonunits in $\operatorname{End}_R(M)$ are additively closed. Suppose therefore that $f, g \in \operatorname{End}_R(M) \setminus \operatorname{End}_R(M)^\times$ are such that $h := f + g \in \operatorname{End}_R(M)^\times$. By multiplying by h^{-1} we reduce to the case that $h = \mathrm{id}_M$. Then the left-hand side in the identity

$$g = \mathrm{id}_M - f$$

is not a unit, but the right-hand side is by what we have shown above. This is a contradiction, and hence $\operatorname{End}_R(M) \setminus \operatorname{End}_R(M)^\times$ is additively closed. □

Proposition 1.4.6 *Let*

$$M = M_1 \oplus \cdots \oplus M_r = N_1 \oplus \cdots \oplus N_s$$

be two decompositions of the R-module M into indecomposable R-modules M_i and N_j; if $\operatorname{End}_R(M_i)$, for any $1 \le i \le r$, is a local ring we have:

i. $r = s$;
ii. *there is a permutation* σ *of* $\{1, \dots, r\}$ *such that* $N_j \cong M_{\sigma(j)}$ *for any* $1 \le j \le r$.

Proof The proof is by induction with respect to r. The case $r = 1$ is trivial since M then is indecomposable. In general we have the R-module homomorphisms

$$f_i\colon\ M \xrightarrow{\mathrm{pr}_{M_i}} M_i \xrightarrow{\subseteq} M \quad\text{and}\quad g_j\colon\ M \xrightarrow{\mathrm{pr}_{N_j}} N_j \xrightarrow{\subseteq} M$$

in $\mathrm{End}_R(M)$ satisfying the equation

$$\mathrm{id}_M = f_1 + \cdots + f_r = g_1 + \cdots + g_s.$$

In particular, $f_1 = f_1 g_1 + \cdots + f_1 g_s$, which restricts to the equation

$$\mathrm{id}_{M_1} = (\mathrm{pr}_{M_1} g_1)|M_1 + \cdots + (\mathrm{pr}_{M_1} g_s)|M_1$$

in $\mathrm{End}_R(M_1)$. But $\mathrm{End}_R(M_1)$ is local. Hence at least one of the summands must be a unit. By renumbering we may assume that the composed map

$$M_1 \xrightarrow{\subseteq} M \xrightarrow{\mathrm{pr}_{N_1}} N_1 \xrightarrow{\subseteq} M \xrightarrow{\mathrm{pr}_{M_1}} M_1$$

is an automorphism of M_1. This implies that

$$N_1 \cong M_1 \oplus \ker(\mathrm{pr}_{M_1}|N_1).$$

Since N_1 is indecomposable we obtain that $g_1 : M_1 \xrightarrow{\cong} N_1$ is an isomorphism. In particular

$$M_1 \cap \ker(g_1) = M_1 \cap (N_2 \oplus \cdots \oplus N_s) = \{0\}.$$

On the other hand let $x \in N_1$ and write $x = g_1(y)$ with $y \in M_1$. Then

$$g_1(x - y) = x - g_1(y) = 0$$

and hence

$$x = y + (x - y) \in M_1 + \ker(g_1) = M_1 + (N_2 \oplus \cdots \oplus N_s).$$

This shows that $N_1 \subseteq M_1 + (N_2 \oplus \cdots \oplus N_s)$, and hence

$$M = N_1 + \cdots + N_s = M_1 + (N_2 \oplus \cdots \oplus N_s).$$

Together we obtain

$$M = M_1 \oplus N_2 \oplus \cdots \oplus N_s$$

and therefore

$$M/M_1 \cong M_2 \oplus \cdots \oplus M_r \cong N_2 \oplus \cdots \oplus N_s.$$

We now apply the induction hypothesis to these two decompositions of the R-module M/M_1. □

Theorem 1.4.7 (Krull–Remak–Schmidt) *The assumptions of Proposition* 1.4.6 *are satisfied in any of the following cases:*

i. *M is of finite length;*
ii. *R is left artinian and M is finitely generated;*
iii. *R is left noetherian, $R/\operatorname{Jac}(R)$ is left artinian, any finitely generated R-module is complete, and M is finitely generated;*
iv. *R is an R_0-algebra, which is finitely generated as an R_0-module, over a noetherian complete commutative ring R_0 such that $R_0/\operatorname{Jac}(R_0)$ is artinian, and M is finitely generated.*

Proof i. Use Proposition 1.4.4.
ii. This reduces to i. by Proposition 1.1.1.iii.
iii. Use Proposition 1.4.5.
iv. This reduces to iii. by Proposition 1.3.6. □

Example Let K be a field and G be a finite group. The group ring $K[G]$ is left artinian. Hence Proposition 1.4.6 applies: Any finitely generated $K[G]$-module has a "unique" decomposition into indecomposable modules.

1.5 Idempotents and Blocks

Of primary interest to us is the decomposition of the ring R itself into indecomposable submodules. This is closely connected to the existence of idempotents in R.

Definition

i. An element $e \in R$ is called an idempotent if $e^2 = e \neq 0$.
ii. Two idempotents $e_1, e_2 \in R$ are called orthogonal if $e_1 e_2 = 0 = e_2 e_1$.
ii. An idempotent $e \in R$ is called primitive if e is not equal to the sum of two orthogonal idempotents.
iv. The idempotents in the center $Z(R)$ of R are called central idempotents in R.

We note that eRe, for any idempotent $e \in R$, is a subring of R with unit element e. We also note that for any idempotent $1 \neq e \in R$ the element $1 - e$ is another idempotent, and $e, 1 - e$ are orthogonal.

Exercise If $e_1, \ldots, e_r \in R$ are idempotents which are pairwise orthogonal then $e_1 + \cdots + e_r$ is an idempotent as well.

Proposition 1.5.1 *Let $L = Re \subseteq R$ be a left ideal generated by an idempotent e; the map*

$$\begin{array}{ccc} \text{set of all sets } \{e_1, \dots, e_r\} \text{ of pair-} & & \text{set of all decompositions} \\ \text{wise orthogonal idempotents } e_i \in R & \xrightarrow{\sim} & L = L_1 \oplus \cdots \oplus L_r \text{ of } L \\ \text{such that } e_1 + \cdots + e_r = e & & \text{into nonzero left ideals } L_i \\ \{e_1, \dots, e_r\} & \longmapsto & L = Re_1 \oplus \cdots \oplus Re_r \end{array}$$

is bijective.

Proof Suppose given a set $\{e_1, \dots, e_r\}$ in the left-hand side. We have $L = Re = R(e_1 + \cdots + e_r) \subseteq Re_1 + \cdots + Re_r$. On the other hand $L \supseteq Re_i e = R(e_i e_1 + \cdots + e_i e_r) = Re_i$. Hence $L = Re_1 + \cdots + Re_r$. To see that the sum is direct let

$$ae_j = \sum_{i \neq j} a_i e_i \in Re_j \cap \left(\sum_{i \neq j} Re_i \right) \quad \text{with } a, a_i \in R$$

be an arbitrary element. Then

$$ae_j = (ae_j)e_j = \left(\sum_{i \neq j} a_i e_i \right) e_j = \sum_{i \neq j} a_i e_i e_j = 0.$$

It follows that the asserted map is well defined.

To establish its injectivity let $\{e'_1, \dots, e'_r\}$ be another set in the left-hand side such that the two decompositions

$$Re_1 \oplus \cdots \oplus Re_r = L = Re'_1 \oplus \cdots \oplus Re'_r$$

coincide. This means that there is a permutation σ of $\{1, \dots, r\}$ such that $Re'_i = Re_{\sigma(i)}$ for any i. The identity

$$e_{\sigma(1)} + \cdots + e_{\sigma(r)} = e = e'_1 + \cdots + e'_r$$

then implies that $e'_i = e_{\sigma(i)}$ for any i or, equivalently, that $\{e'_1, \dots, e'_r\} = \{e_1, \dots, e_r\}$.

We prove the surjectivity of the asserted map in two steps. *Step 1:* We assume that $e = 1$ and hence $L = R$. Suppose that $R = L_1 \oplus \cdots \oplus L_r$ is a decomposition as a direct sum of nonzero left ideals L_i. We then have $1 = e_1 + \cdots + e_r$ for appropriate elements $e_i \in L_i$ and, in particular, $Re_i \subseteq L_i$ and $R = R \cdot 1 = R(e_1 + \cdots + e_r) \subseteq Re_1 + \cdots + Re_r$. It follows that $Re_i = L_i$. Since $L_i \neq \{0\}$ we must have $e_i \neq 0$. Furthermore,

$$e_i = e_i \cdot 1 = e_i(e_1 + \cdots + e_r) = e_i e_1 + \cdots + e_i e_r \quad \text{for any } 1 \leq i \leq r.$$

Since $e_i e_j \in L_j$ we obtain

$$e_i^2 = e_i \quad \text{and} \quad e_i e_j = 0 \quad \text{for } j \neq i.$$

We conclude that the set $\{e_1, \dots, e_r\}$ is a preimage of the given decomposition under the map in the assertion. *Step 2:* For a general $e \neq 1$ we first observe that the elements $1 - e$ and e are orthogonal idempotents in R. Hence, by what we have shown already, we have the decomposition $R = R(1-e) \oplus Re = R(1-e) \oplus L$. Suppose now that $L = L_1 \oplus \cdots \oplus L_r$ is a direct sum decomposition into nonzero left ideals L_i. Then

$$R = R(1-e) \oplus L_1 \oplus \cdots \oplus L_r$$

is a decomposition into nonzero left ideals as well. By Step 1 we find pairwise orthogonal idempotents $e_0, e_1, \dots, e_r$ in R such that

$$1 = e_0 + e_1 + \cdots + e_r, \qquad Re_0 = R(1-e), \quad \text{and} \quad Re_i = L_i \quad \text{for } 1 \leq i \leq r.$$

Comparing this with the identity $1 = (1-e) + e$ where $1 - e \in Re_0$ and $e \in L = L_1 \oplus \cdots \oplus L_r$ we see that $1 - e = e_0$ and $e = e_1 + \cdots + e_r$. Therefore the set $\{e_1, \dots, e_r\}$ is a preimage of the decomposition $L = L_1 \oplus \cdots \oplus L_r$ under the asserted map. □

There is a completely analogous right ideal version, sending $\{e_1, \dots, e_r\}$ to $e_1 R \oplus \cdots \oplus e_r R$, of the above proposition.

Corollary 1.5.2 *For any idempotent $e \in R$ the following conditions are equivalent:*

i. *The R-module Re is indecomposable;*
ii. *e is primitive;*
iii. *the right R-module eR is indecomposable;*
iv. *the ring eRe contains no idempotent other than e.*

Proof The equivalence of i., ii., and iii. follows immediately from Proposition 1.5.1 and its right ideal version.

ii. $\Longrightarrow$ iv. Suppose that $e \neq f \in eRe$ is an idempotent. Then $ef = fe = f$, and $e = (e - f) + f$ is the sum of the orthogonal idempotents $e - f$ and f. This is a contradiction to the primitivity of e.

iv. $\Longrightarrow$ ii. Suppose that $e = e_1 + e_2$ is the sum of the orthogonal idempotents $e_1, e_2 \in R$. Then $ee_1 = e_1^2 + e_2 e_1 = e_1$ and $e_1 e = e_1^2 + e_1 e_2 = e_1$ and hence $e_1 = ee_1 e \in eRe$. Since $e_1 \neq e$ this again is a contradiction. □

Proposition 1.5.3 *Let $I = Re = eR$ be a two-sided ideal generated by a central idempotent e; we then have:*

i. *I is a subring of R with unit element e;*

ii. *the map*

$$\begin{array}{ccc} \text{set of all sets } \{e_1, \ldots, e_r\} & & \text{set of all decompositions} \\ \text{of pairwise orthogonal} & \xrightarrow{\sim} & I = I_1 \oplus \cdots \oplus I_r \\ \text{idempotents } e_i \in Z(R) \text{ such that} & & \text{of } I \text{ into nonzero two-sided} \\ e_1 + \cdots + e_r = e & & \text{ideals } I_i \subseteq R \\ \{e_1, \ldots, e_r\} & \longmapsto & I = Re_1 \oplus \cdots \oplus Re_r \end{array}$$

is bijective; moreover, the multiplication in $I = Re_1 \oplus \cdots \oplus Re_r$ *can be carried out componentwise.*

Proof i. We have $I = eRe$. ii. Obviously, since e_i is central the ideal $Re_i = e_i R$ is two-sided. Taking Proposition 1.5.1 into account it therefore remains to show that, if in the decomposition $I = Re_1 \oplus \cdots \oplus Re_r$ with $e = e_1 + \cdots + e_r$ the Re_i are two-sided ideals, then the e_i necessarily are central. But for any $a \in R$ we have

$$\begin{aligned} ae_1 + \cdots + ae_r = ae = ea &= a(e_1 + \cdots + e_r) \\ &= (e_1 + \cdots + e_r)a \\ &= e_1 a + \cdots + e_r a \end{aligned}$$

where ae_i and $e_i a$ both lie in Re_i. Hence $ae_i = e_i a$ since the summands are uniquely determined in Re_i. For the second part of the assertion let $a = a_1 + \cdots + a_r$ and $b = b_1 + \cdots + b_r$ with $a_i, b_i \in I_i$ be arbitrary elements. Then $a_i = a_i e_i$ and $b_i = e_i b_i$ and hence

$$\begin{aligned} ab &= (a_1 + \cdots + a_r)(b_1 + \cdots + b_r) \\ &= (a_1 e_1 + \cdots + a_r e_r)(e_1 b_1 + \cdots + e_r b_r) \\ &= \sum_{i,j} a_i e_i e_j b_j = \sum_i a_i e_i b_i \\ &= \sum_i a_i b_i. \end{aligned}$$

□

Corollary 1.5.4 *A central idempotent* $e \in R$ *is primitive in* $Z(R)$ *if and only if* Re *is not the direct sum of two nonzero two-sided ideals of* R.

Proposition 1.5.5 *If* R *is left noetherian then we have*:

i. $1 \in R$ *can be written as a sum of pairwise orthogonal primitive idempotents*;
ii. R *contains only finitely many central idempotents*;
iii. *any two different central idempotents which are primitive in* $Z(R)$ *are orthogonal*;

iv. *if* $\{e_1, \ldots, e_n\}$ *is the set of all central idempotents which are primitive in* $Z(R)$ *then* $e_1 + \cdots + e_n = 1$.

Proof i. By Lemma 1.1.6 we have a direct sum decomposition $R = L_1 \oplus \cdots \oplus L_r$ of R into indecomposable left ideals L_i. According to Proposition 1.5.1 this corresponds to a decomposition of the unit element

$$1 = f_1 + \cdots + f_r \tag{1.5.1}$$

as a sum of pairwise orthogonal idempotents f_i such that $L_i = Rf_i$. Moreover, Corollary 1.5.2 says that each f_i is primitive.

ii. We keep the decomposition (1.5.1). Let $e \in Z(R)$ be any idempotent. Then $e = ef_1 + \cdots + ef_r$ and $f_i = ef_i + (1-e)f_i$. We have

$$(ef_i)^2 = e^2 f_i^2 = ef_i, \qquad ((1-e)f_i)^2 = (1-e)^2 f_i^2 = (1-e)f_i,$$
$$ef_i(1-e)f_i = e(1-e)f_i^2 = 0, \quad \text{and} \quad (1-e)f_i ef_i = (1-e)ef_i^2 = 0.$$

But f_i is primitive. Hence either $ef_i = 0$ or $(1-e)f_i = 0$, i.e. $ef_i = f_i$. This shows that there is a subset $S \subseteq \{1, \ldots, r\}$ such that

$$e = \sum_{i \in S} f_i,$$

and we see that there are only finitely many possibilities for e.

iii. Let $e_1 \neq e_2$ be two primitive idempotents in the ring $Z(R)$. We then have

$$e_1 = e_1 e_2 + e_1(1 - e_2)$$

where the summands satisfy $(e_1e_2)^2 = e_1e_2$, $(e_1(1-e_2))^2 = e_1(1-e_2)$, and $e_1e_2e_1(1-e_2) = 0$. Hence $e_1e_2 = 0$ or $e_1 = e_1e_2$. By symmetry we also obtain $e_2e_1 = 0$ or $e_2 = e_2e_1$. It follows that $e_1e_2 = 0$ or $e_1 = e_1e_2 = e_2$. The latter case being excluded by assumption we conclude that $e_1e_2 = 0$.

iv. First we consider any idempotent $f \in Z(R)$. If f is not primitive in $Z(R)$ then we can write it as the sum of two orthogonal idempotents in $Z(R)$. Any of the two summands either is primitive or again can be written as the sum of two new (exercise!) orthogonal idempotents in $Z(R)$. Proceeding in this way we must arrive, because of ii., after finitely many steps at an expression of f as a sum of pairwise orthogonal idempotents which are primitive in $Z(R)$. This, first of all, shows that the set $\{e_1, \ldots, e_n\}$ is nonempty. By iii. the sum $e := e_1 + \cdots + e_n$ is a central idempotent. Suppose that $e \neq 1$. Then $1 - e$ is a central idempotent. By the initial observation we find a subset $S \subseteq \{1, \ldots, n\}$ such that

$$1 - e = \sum_{i \in S} e_i.$$

For $i \in S$ we then compute

$$e_i = e_i\left(\sum_{j\in S} e_j\right) = e_i(1-e) = e_i - e_i(e_1 + \cdots + e_n) = e_i - e_i = 0$$

which is a contradiction. □

Let $e \in R$ be a central idempotent which is primitive in $Z(R)$. An R-module M is said to belong to the *e-block* of R if $eM = M$ holds true.

Exercise If M belongs to the e-block then we have:

a. $ex = x$ for any $x \in M$;
b. every submodule and every factor module of M also belongs to the e-block.

We now suppose that R is left noetherian. Let $\{e_1, \ldots, e_n\}$ be the set of all central idempotents which are primitive in $Z(R)$. By Proposition 1.5.5.iii/iv the e_i are pairwise orthogonal and satisfy $e_1 + \cdots + e_n = 1$. Let M be any R-module. Since e_i is central $e_iM \subseteq M$ is a submodule which obviously belongs to the e_i-block. We also have $M = 1 \cdot M = (e_1 + \cdots + e_n)M \subseteq e_1M + \cdots + e_nM$ and hence $M = e_1M + \cdots + e_nM$. For

$$e_ix = \sum_{j\neq i} e_jx_j \in e_iM \cap \left(\sum_{j\neq i} e_jM\right) \quad \text{with } x, x_j \in M$$

we compute

$$e_ix = e_ie_ix = e_i\sum_{j\neq i} e_jx_j = \sum_{j\neq i} e_ie_jx_j = 0.$$

Hence the decomposition

$$M = e_1M \oplus \cdots \oplus e_nM$$

is direct. It is called the *block decomposition* of M.

Remark 1.5.6

i. For any R-module homomorphism $f : M \longrightarrow N$ we have $f(e_iM) \subseteq e_iN$.
ii. If the submodule $N \subseteq M$ lies in the e_i-block then $N \subseteq e_iM$.
iii. If M is indecomposable then there is a unique $1 \leq i \leq n$ such that M lies in the e_i-block.

Proof i. Since f is R-linear we have $f(e_iM) = e_if(M) \subseteq e_iN$. ii. Apply i. to the inclusion $e_iN = N \subseteq M$. iii. In this case the block decomposition can have only a single nonzero summand. □

As a consequence of Proposition 1.5.3 the map

$$R \xrightarrow{\cong} \prod_{i=1}^{n} Re_i$$
$$a \longmapsto (ae_1, \ldots, ae_n)$$

is an isomorphism of rings. If M lies in the e_i-block then

$$ax = a(e_i x) = (ae_i)x \quad \text{for any } a \in R \text{ and } x \in M.$$

This means that M comes, by restriction of scalars along the projection map $R \longrightarrow Re_i$, from an Re_i-module (with the same underlying additive group). In this sense the e_i-block coincides with the class of all Re_i-modules.

We next discuss, for arbitrary R, the relationship between idempotents in the ring R and in a factor ring R/I.

Proposition 1.5.7 *Let $I \subseteq R$ be a two-sided ideal and suppose that either every element in I is nilpotent or R is I-adically complete; then for any idempotent $\varepsilon \in R/I$ there is an idempotent $e \in R$ such that $e + I = \varepsilon$.*

Proof Case 1: We assume that every element in I is nilpotent. Let $\varepsilon = a + I$ and put $b := 1 - a$. Then $ab = ba = a - a^2 \in I$, and hence $(ab)^m = 0$ for some $m \geq 1$. Since a and b commute the binomial theorem gives

$$1 = (a+b)^{2m} = \sum_{i=0}^{2m} \binom{2m}{i} a^{2m-i} b^i = e + f$$

with

$$e := \sum_{i=0}^{m} \binom{2m}{i} a^{2m-i} b^i \in aR \quad \text{and} \quad f := \sum_{j=m+1}^{2m} \binom{2m}{j} a^{2m-j} b^j.$$

For any $0 \leq i \leq m$ and $m < j \leq 2m$ we have

$$a^{2m-i} b^i a^{2m-j} b^j = a^m b^m a^{3m-i-j} b^{i+j-m} = (ab)^m a^{3m-i-j} b^{i+j-m} = 0$$

and hence $ef = 0$. It follows that $e = e(e+f) = e^2$. Moreover, $ab \in I$ implies

$$e + I = a^{2m} + \left(\sum_{i=1}^{m} \binom{2m}{i} a^{2m-i-1} b^{i-1} \right) ab + I = a^{2m} + I = \varepsilon^{2m} = \varepsilon.$$

Case 2: We assume that R is I-adically complete. Since $R \xrightarrow{\cong} \varprojlim R/I^n$ it suffices to construct a sequence of idempotents $\varepsilon_n \in R/I^n$, for $n \geq 2$, such that $\mathrm{pr}(\varepsilon_{n+1}) = \varepsilon_n$ and $\varepsilon_1 = \varepsilon$. Because of $I^{2n} \subseteq I^{n+1}$ the ideal I^n/I^{n+1} in the ring R/I^{n+1} is nilpotent. Hence we may, inductively, apply the first case to the idempotent ε_n in the factor ring R/I^n of the ring R/I^{n+1} in order to obtain ε_{n+1}. □

We point out that, by Proposition 1.2.1.v and Lemma 1.3.2, the assumptions in Proposition 1.5.7 imply that $I \subseteq \mathrm{Jac}(R)$.

Remark 1.5.8

i. $\mathrm{Jac}(R)$ does not contain any idempotent.
ii. Let $I \subseteq \mathrm{Jac}(R)$ be a two-sided ideal and $e \in R$ be an idempotent; we then have:
   a. $e + I$ is an idempotent in R/I;
   b. if $e + I \in R/I$ is primitive then e is primitive.

Proof i. Suppose that $e \in \mathrm{Jac}(R)$ is an idempotent. Then $1 - e$ is an idempotent as well as a unit (cf. Proposition 1.2.1.iv). Multiplying the equation $(1-e)^2 = 1 - e$ by the inverse of $1 - e$ shows that $1 - e = 1$. This leads to the contradiction that $e = 0$.

ii. The assertion a. is immediate from i. For b. let $e = e_1 + e_2$ with orthogonal idempotents $e_1, e_2 \in R$. Then $e + I = (e_1 + I) + (e_2 + I)$ with $(e_1 + I)(e_2 + I) = e_1 e_2 + I = I$. Therefore, by a., $e_1 + I$ and $e_2 + I$ are orthogonal idempotents in R/I. This, again, is a contradiction. □

Lemma 1.5.9

i. *Let $e, f \in R$ be two idempotents such that $e + \mathrm{Jac}(R) = f + \mathrm{Jac}(R)$; then $Re \cong Rf$ as R-modules.*
ii. *Let $I \subseteq \mathrm{Jac}(R)$ be a two-sided ideal and $e, f \in R$ be idempotents such that the idempotents $e + I, f + I \in R/I$ are orthogonal; then there exists an idempotent $f' \in R$ such that $f' + I = f + I$ and e, f' are orthogonal.*

Proof i. We consider the pair of R-modules $L := Rfe \subseteq M := Re$. Since $f - e \in \mathrm{Jac}(R)$ we have

$$L + \mathrm{Jac}(R)M = Rfe + \mathrm{Jac}(R)e = R\big(e + (f-e)\big)e + \mathrm{Jac}(R)e$$
$$= Re + \mathrm{Jac}(R)e = Re = M.$$

The factor module M/L being generated by a single element $e + L$ we may apply the Nakayama lemma 1.2.3 and obtain $Rfe = Re$. On the other hand the decomposition $R = Rf \oplus R(1-f)$ leads to $Re = Rfe \oplus R(1-f)e = Re \oplus R(1-f)e$. It follows that $R(1-f)e = \{0\}$ and, in particular, $(1-f)e = 0$. We obtain that

$$e = fe$$

and, by symmetry, also

$$f = ef.$$

This easily implies that the R-module homomorphisms of right multiplication by f and e, respectively,

$$\begin{aligned} Re &\rightleftarrows Rf \\ re &\mapsto ref \\ r'fe &\leftarrow\!\shortmid r'f \end{aligned}$$

are inverse to each other.

ii. We have $fe \in I \subseteq \mathrm{Jac}(R)$, hence $1 - fe \in R^\times$, and we may introduce the idempotent

$$f_0 := (1 - fe)^{-1} f (1 - fe).$$

Obviously $f_0 + I = f + I$ and $f_0 e = (1 - fe)^{-1} f(e - fe) = (1 - fe)^{-1}(fe - fe) = 0$. We put

$$f' := (1 - e) f_0.$$

Then

$$f' + I = f_0 - ef_0 + I = (f + I) - (e + I)(f + I) = f + I.$$

Moreover,

$$f'e = (1 - e) f_0 e = 0 \quad \text{and} \quad ef' = e(1 - e) f_0 = 0.$$

Finally

$$f'^2 = (1 - e) f_0 (1 - e) f_0 = (1 - e)\big(f_0^2 - f_0 e f_0\big) = (1 - e) f_0 = f'.$$

□

Proposition 1.5.10 *Under the assumptions of Proposition* 1.5.7 *we have:*

i. *If $e \in R$ is a primitive idempotent then the idempotent $e + I \in R/I$ is primitive as well;*
ii. *if $\varepsilon_1, \ldots, \varepsilon_r \in R/I$ are pairwise orthogonal idempotents then there are pairwise orthogonal idempotents $e_1, \ldots, e_r \in R$ such that $\varepsilon_i = e_i + I$ for any $1 \le i \le r$.*

Proof i. Let $e + I = \varepsilon_1 + \varepsilon_2$ with orthogonal idempotents $\varepsilon_1, \varepsilon_2 \in R/I$. By Proposition 1.5.7 we find idempotents $e_i \in R$ such that $\varepsilon_i = e_i + I$, for $i = 1, 2$. By Lemma 1.5.9.ii there is an idempotent $e_2' \in R$ such that $e_2' + I = e_2 + I = \varepsilon_2$ and e_1, e_2' are orthogonal. The latter implies that $f := e_1 + e_2'$ is an idempotent as well. It satisfies $f + I = \varepsilon_1 + \varepsilon_2 = e + I$. Hence $Rf \cong Re$ by Lemma 1.5.9.i. Applying Proposition 1.5.1 we obtain that $Re_1 \oplus Re_2' \cong Re$ and we see that e is not primitive. This is a contradiction.

ii. The proof is by induction with respect to r. We assume that the idempotents $e_1, \ldots, e_{r-1}$ have been constructed already. On the one hand we then have the idempotent $e := e_1 + \cdots + e_{r-1}$. On the other hand we find, by Proposition 1.5.7, an idempotent $f \in R$ such that $f + I = \varepsilon_r$. Since $e + I = \varepsilon_1 + \cdots + \varepsilon_{r-1}$ and ε_r are orthogonal there exists, by Lemma 1.5.9.ii, an idempotent $e_r \in R$ such that $e_r + I = f + I = \varepsilon_r$ and e, e_r are orthogonal. It remains to observe that

$$e_r e_i = e_r(ee_i) = (e_r e)e_i = 0 \quad \text{and} \quad e_i e_r = e_i e e_r = 0$$

for any $1 \leq i < r$. □

Proposition 1.5.11 *Suppose that R is complete and that $R/\operatorname{Jac}(R)$ is left artinian; then R is local if and only if 1 is the only idempotent in R.*

Proof We first assume that R is local. Let $e \in R$ be any idempotent. Since $e \notin \operatorname{Jac}(R)$ by Remark 1.5.8.i we must have $e \in R^\times$. Multiplying the identity $e^2 = e$ by e^{-1} gives $e = 1$. Now let us assume, vice versa, that 1 is the only idempotent in R. As a consequence of Proposition 1.5.7, the factor ring $\overline{R} := R/\operatorname{Jac}(R)$ also has no other idempotent than 1. Therefore, by Proposition 1.5.1, the $\overline{R}$-module $L := \overline{R}$ is indecomposable. On the other hand, the ring $\overline{R}$ being left artinian the module L, by Corollary 1.2.2, is of finite length. Hence Proposition 1.4.4 implies that $\operatorname{End}_{\overline{R}}(L)$ is a local ring. But the map

$$\begin{aligned} \overline{R}^{\text{op}} &\xrightarrow{\cong} \operatorname{End}_{\overline{R}}(\overline{R}) \\ c &\longmapsto [a \mapsto ac] \end{aligned}$$

is an isomorphism of rings (exercise!). We obtain that $\overline{R}^{\text{op}}$ and $\overline{R}$ are local rings. Since $\operatorname{Jac}(\overline{R}) = \operatorname{Jac}(R/\operatorname{Jac}(R)) = \{0\}$ the ring $\overline{R}$ in fact is a skew field. Now Proposition 1.4.1 implies that R is local. □

Proposition 1.5.12 *Suppose that R is an R_0-algebra, which is finitely generated as an R_0-module, over a noetherian complete commutative ring R_0 such that $R_0/\operatorname{Jac}(R_0)$ is artinian; then the map*

set of all central idempotents in R	$\xrightarrow{\cong}$	*set of all central idempotents in $R/\operatorname{Jac}(R_0)R$*
e	$\longmapsto$	$\overline{e} := e + \operatorname{Jac}(R_0)R$

is bijective; moreover, this bijection satisfies:

- *e, f are orthogonal if and only if $\overline{e}, \overline{f}$ are orthogonal;*
- *e is primitive in $Z(R)$ if and only if $\overline{e}$ is primitive in $Z(R/\operatorname{Jac}(R_0)R)$.*

Proof In Lemma 1.3.5.iii we have seen that $\operatorname{Jac}(R_0)R \subseteq \operatorname{Jac}(R)$. Hence $\operatorname{Jac}(R_0)R$, by Remark 1.5.8.i, does not contain any idempotent. Therefore $\overline{e} \neq 0$ which says that the map in the assertion is well defined. To establish its injectivity let us assume that $\overline{e_1} = \overline{e_2}$. Then

$$\overline{e_i} - \overline{e_1}\,\overline{e_2} = 0 \quad \text{and hence} \quad e_i - e_1 e_2 \in \operatorname{Jac}(R_0)R.$$

But since e_1 and e_2 commute we have

$$(e_i - e_1e_2)^2 = e_i^2 - 2e_ie_1e_2 + e_1^2e_2^2 = e_i - 2e_1e_2 + e_1e_2 = e_i - e_1e_2.$$

It follows that $e_i - e_1e_2 = 0$ which implies $e_1 = e_1e_2 = e_2$.

For the surjectivity let $\varepsilon \in \overline{R} := R/\mathrm{Jac}(R_0)R$ be any central idempotent. In the proof of Proposition 1.3.6 we have seen that R is $\mathrm{Jac}(R_0)R$-adically complete. Hence we may apply Proposition 1.5.7 and obtain an idempotent $e \in R$ such that $\overline{e} = \varepsilon$. We, in fact, claim the stronger statement that any such e necessarily lies in the center $Z(R)$ of R. We have

$$R = Re + R(1-e) = eRe + (1-e)Re + eR(1-e) + (1-e)R(1-e)$$

and correspondingly

$$\overline{R} = \varepsilon\overline{R}\varepsilon + (1-\varepsilon)\overline{R}\varepsilon + \varepsilon\overline{R}(1-\varepsilon) + (1-\varepsilon)\overline{R}(1-\varepsilon).$$

But

$$(1-\varepsilon)\overline{R}\varepsilon = (1-\varepsilon)\varepsilon\overline{R} = \{0\} \quad \text{and} \quad \varepsilon\overline{R}(1-\varepsilon) = \overline{R}\varepsilon(1-\varepsilon) = \{0\}$$

since ε is central in $\overline{R}$. It follows that $(1-e)Re$ and $eR(1-e)$ both are contained in $\mathrm{Jac}(R_0)R$. We see that

$$(1-e)Re = (1-e)^2Re^2 \subseteq (1-e)\,\mathrm{Jac}(R_0)Re = \mathrm{Jac}(R_0)(1-e)Re$$

and similarly $eR(1-e) \subseteq \mathrm{Jac}(R_0)eR(1-e)$. This means that for the two finitely generated (as submodules of R) R_0-modules $M := (1-e)Re$ and $M := eR(1-e)$ we have $\mathrm{Jac}(R_0)M = M$. The Nakayama lemma 1.2.3 therefore implies $(1-e)Re = eR(1-e) = \{0\}$ and consequently that

$$R = eRe + (1-e)R(1-e).$$

If we write an arbitrary element $a \in R$ as $a = ebe + (1-e)c(1-e)$ with $b, c \in R$ then we obtain

$$ea = e(ebe) = ebe = (ebe)e = ae.$$

This shows that $e \in Z(R)$.

For the second half of the assertion we first note that with e, f also $\overline{e}, \overline{f}$ are orthogonal for trivial reasons. Let us suppose, vice versa, that $\overline{e}, \overline{f}$ are orthogonal. By Lemma 1.5.9.ii we find an idempotent $f' \in R$ such that $\overline{f'} = \overline{f}$ and e, f' are orthogonal. According to what we have established above f' necessarily is central. Hence the injectivity of our map forces $f' = f$. So e, f are orthogonal.

Finally, if $e = e_1 + e_2$ with orthogonal central idempotents e_1, e_2 then obviously $\overline{e} = \overline{e_1} + \overline{e_2}$ with orthogonal idempotents $\overline{e_1}, \overline{e_2} \in Z(\overline{R})$. Vice versa, let $\overline{e} = \varepsilon_1 + \varepsilon_2$ with orthogonal idempotents $\varepsilon_1, \varepsilon_2 \in Z(\overline{R})$. By the surjectivity of our map we find idempotents $e_i \in Z(R)$ such that $\overline{e_i} = \varepsilon_i$. We have shown already that e_1, e_2 necessarily are orthogonal. Since $e_1 + e_2$ then is a central idempotent with $\overline{e_1 + e_2} = \overline{e_1} + \overline{e_2} = \varepsilon_1 + \varepsilon_2 = \overline{e}$ the injectivity of our map forces $e_1 + e_2 = e$. □

Remark 1.5.13 Let $e \in R$ be any idempotent, and let $L \subseteq M$ be R-modules; then $eM/eL \cong e(M/L)$ as $Z(R)$-modules.

Proof We have the obviously well-defined and surjective $Z(R)$-module homomorphism

$$\begin{aligned} eM/eL &\longrightarrow e(M/L) \\ ex + eL &\longmapsto e(x+L) = ex + L. \end{aligned}$$

If $ex + L = L$ then $ex \in L$ and hence $ex = e(ex) \in eL$. This shows that the map is injective as well. □

Keeping the assumptions of Proposition 1.5.12 we consider the block decomposition

$$M = e_1 M \oplus \cdots \oplus e_n M$$

of any R-module M. It follows that

$$\begin{aligned} M/\operatorname{Jac}(R_0)M &= e_1 M/\operatorname{Jac}(R_0)e_1 M \oplus \cdots \oplus e_n M/\operatorname{Jac}(R_0)e_n M \\ &= e_1 M/e_1 \operatorname{Jac}(R_0)M \oplus \cdots \oplus e_n M/e_n \operatorname{Jac}(R_0)M \\ &= \overline{e_1}\big(M/\operatorname{Jac}(R_0)M\big) \oplus \cdots \oplus \overline{e_n}\big(M/\operatorname{Jac}(R_0)M\big) \end{aligned}$$

is the block decomposition of the $R/\operatorname{Jac}(R_0)R$-module $M/\operatorname{Jac}(R_0)M$. If $e_i M = \{0\}$ then obviously $\overline{e_i}(M/\operatorname{Jac}(R_0)M) = \{0\}$. Vice versa, let us suppose that $\overline{e_i}(M/\operatorname{Jac}(R_0)M) = \{0\}$. Then $e_i M \subseteq \operatorname{Jac}(R_0)M$ and hence

$$e_i M = e_i^2 M \subseteq \operatorname{Jac}(R_0)e_i M.$$

If M is finitely generated as an R-module then $e_i M$ is finitely generated as an R_0-module and the Nakayama lemma 1.2.3 implies that $e_i M = \{0\}$. We, in particular, obtain that a finitely generated R-module M belongs to the e_i-block if and only if $M/\operatorname{Jac}(R_0)M$ belongs to the $\overline{e_i}$-block.

1.6 Projective Modules

We fix an R-module X. For any R-module M, resp. for any R-module homomorphism $g : L \longrightarrow M$, we have the $Z(R)$-module

$$\operatorname{Hom}_R(X, M),$$

resp. the $Z(R)$-module homomorphism

$$\begin{aligned} \operatorname{Hom}_R(X, g)\colon \quad \operatorname{Hom}_R(X, L) &\longrightarrow \operatorname{Hom}_R(X, M) \\ f &\longmapsto g \circ f. \end{aligned}$$

Lemma 1.6.1 *For any exact sequence* $0 \longrightarrow L \xrightarrow{h} M \xrightarrow{g} N$ *of R-modules the sequence*

$$0 \longrightarrow \mathrm{Hom}_R(X, L) \xrightarrow{\mathrm{Hom}_R(X,h)} \mathrm{Hom}_R(X, M) \xrightarrow{\mathrm{Hom}_R(X,g)} \mathrm{Hom}_R(X, N)$$

is exact as well.

Proof Whenever a composite

$$X \xrightarrow{f} L \xrightarrow{h} M$$

is the zero map we must have $f = 0$ since h is injective. This shows the injectivity of $\mathrm{Hom}_R(X, h)$. We have

$$\mathrm{Hom}_R(X, g) \circ \mathrm{Hom}_R(X, h) = \mathrm{Hom}_R(X, g \circ h) = \mathrm{Hom}_R(X, 0) = 0.$$

Hence the image of $\mathrm{Hom}_R(X, h)$ is contained in the kernel of $\mathrm{Hom}_R(X, g)$. Let now $f : X \longrightarrow M$ be an R-module homomorphism such that $g \circ f = 0$. Then $\mathrm{im}(f) \subseteq \ker(g)$. Hence for any $x \in X$ we find, by the exactness of the original sequence, a unique $f_L(x) \in L$ such that

$$f(x) = h\big(f_L(x)\big).$$

This $f_L : X \longrightarrow L$ is an R-module homomorphism such that $h \circ f_L = f$. Hence

$$\text{image of } \mathrm{Hom}_R(X, h) = \text{kernel of } \mathrm{Hom}_R(X, g). \qquad \square$$

Example Let $R = \mathbb{Z}$, $X = \mathbb{Z}/n\mathbb{Z}$ for some $n \geq 2$, and $g : \mathbb{Z} \longrightarrow \mathbb{Z}/n\mathbb{Z}$ be the surjective projection map. Then $\mathrm{Hom}_{\mathbb{Z}}(X, \mathbb{Z}) = \{0\}$ but $\mathrm{Hom}_{\mathbb{Z}}(X, \mathbb{Z}/n\mathbb{Z}) \ni \mathrm{id}_X \neq 0$. Hence the map $\mathrm{Hom}_{\mathbb{Z}}(X, g) : \mathrm{Hom}_{\mathbb{Z}}(X, \mathbb{Z}) \longrightarrow \mathrm{Hom}_{\mathbb{Z}}(X, \mathbb{Z}/n\mathbb{Z})$ cannot be surjective.

Definition An R-module P is called projective if, for any surjective R-module homomorphism $g : M \longrightarrow N$, the map

$$\mathrm{Hom}_R(P, g): \quad \mathrm{Hom}_R(P, M) \longrightarrow \mathrm{Hom}_R(P, N)$$

is surjective.

In slightly more explicit terms, an R-module P is projective if and only if any exact diagram of the form

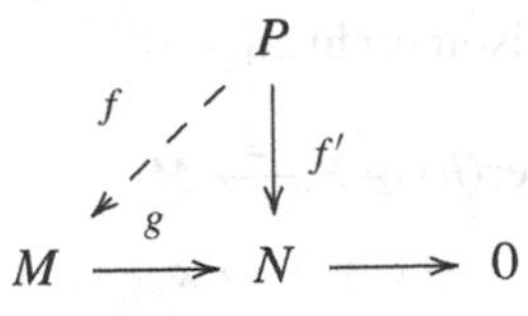

$$\begin{array}{ccccc} & & P & & \\ & \swarrow^{f} & \downarrow f' & & \\ M & \xrightarrow{g} & N & \longrightarrow & 0 \end{array}$$

can be completed, by an oblique arrow f, to a commutative diagram.

Lemma 1.6.2 *For an R-module P the following conditions are equivalent:*

i. *P is projective;*
ii. *for any surjective R-module homomorphism $h : M \longrightarrow P$ there exists an R-module homomorphism $s : P \longrightarrow M$ such that $h \circ s = \mathrm{id}_P$.*

Proof i. $\Longrightarrow$ ii. We obtain s by contemplating the diagram

$$\begin{array}{ccccc} & & P & & \\ & \swarrow^{s} & \downarrow \mathrm{id}_P & & \\ M & \xrightarrow{h} & P & \longrightarrow & 0. \end{array}$$

ii. $\Longrightarrow$ i. Let

$$\begin{array}{ccccc} & & P & & \\ & & \downarrow f' & & \\ M & \xrightarrow{g} & N & \longrightarrow & 0 \end{array}$$

be any exact "test diagram". In the direct sum $M \oplus P$ we have the submodule

$$M' := \{(x, y) \in M \oplus P : g(x) = f'(y)\}.$$

The diagram

$$\begin{array}{ccc} M' & \xrightarrow{h((x,y)):=y} & P \\ {\scriptstyle f''((x,y)):=x} \downarrow & & \downarrow f' \\ M & \xrightarrow{g} & N \end{array}$$

is commutative. We claim that the map h is surjective. Let $y \in P$ be an arbitrary element. Since g is surjective we find an $x \in M$ such that $g(x) = f'(y)$. Then $(x, y) \in M'$ with $h((x, y)) = y$. Hence, by assumption, there is an $s : P \longrightarrow M'$ such that $h \circ s = \mathrm{id}_P$. We define $f := f'' \circ s$ and have

$$g \circ f = g \circ f'' \circ s = f' \circ h \circ s = f' \circ \mathrm{id}_P = f'. \qquad \square$$

In the situation of Lemma 1.6.2.ii we see that P is isomorphic to a direct summand of M via the R-module isomorphism

$$\begin{aligned} \ker(h) \oplus P &\xrightarrow{\cong} M \\ (x, y) &\longmapsto x + s(y) \end{aligned}$$

(exercise!).

Definition An R-module F is called free if there exists an R-module isomorphism

$$F \cong \oplus_{i\in I} R$$

for some index set I.

Example If $R = K$ is a field then, by the existence of bases for vector spaces, any K-module is free. On the other hand for $R = \mathbb{Z}$ the modules $\mathbb{Z}/n\mathbb{Z}$, for any $n \geq 2$, are neither free nor projective.

Lemma 1.6.3 *Any free R-module F is projective.*

Proof If $F \cong \oplus_{i\in I} R$ then let $e_i \in F$ be the element which corresponds to the tuple $(\ldots, 0, 1, 0, \ldots)$ with 1 in the ith place and zeros elsewhere. The set $\{e_i\}_{i\in I}$ is an "R-basis" of the module F. In particular, for any R-module M, the map

$$\begin{aligned} \mathrm{Hom}_R(F, M) &\xrightarrow{\cong} \prod_{i\in I} M \\ f &\longmapsto \big(f(e_i)\big)_i \end{aligned}$$

is bijective. For any surjective R-module homomorphism $M \xrightarrow{g} N$ the lower horizontal map in the commutative diagram

$$\begin{array}{ccc} \mathrm{Hom}_R(F, M) & \xrightarrow{\mathrm{Hom}_R(F,g)} & \mathrm{Hom}_R(F, N) \\ \cong \downarrow & & \downarrow \cong \\ \prod_{i\in I} M & \xrightarrow{(x_i)_i \longmapsto (g(x_i))_i} & \prod_{i\in I} N \end{array}$$

obviously is surjective. Hence the upper one is surjective, too. □

Proposition 1.6.4 *An R-module P is projective if and only if it isomorphic to a direct summand of a free module.*

Proof First we suppose that P is projective. For a sufficiently large index set I we find a surjective R-module homomorphism

$$h\colon \quad F := \bigoplus_{i\in I} R \longrightarrow P.$$

By Lemma 1.6.2 there exists an R-module homomorphism $s : P \longrightarrow F$ such that $h \circ s = \mathrm{id}_P$, and

$$\ker(h) \oplus P \cong F.$$

Vice versa, let P be isomorphic to a direct summand of a free R-module F. Any module isomorphic to a projective module itself is projective (exercise!). Hence we may assume that

$$F = P \oplus P'$$

for some submodule $P' \subseteq F$. We then have the inclusion map $i : P \overset{\subseteq}{\longrightarrow} F$ as well as the projection map $\mathrm{pr} : F \longrightarrow P$. We consider any exact "test diagram"

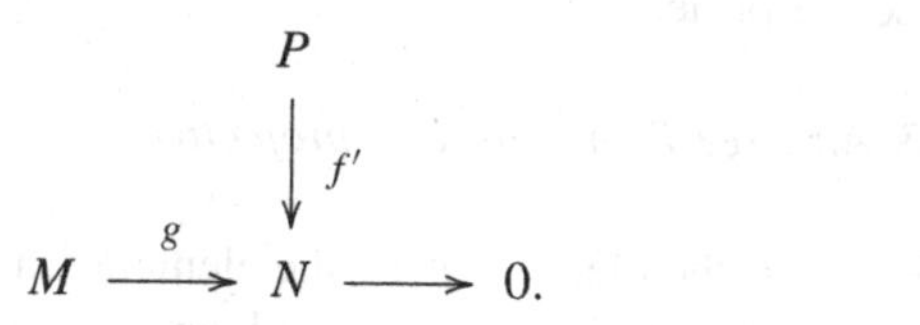

By Lemma 1.6.3 the extended diagram

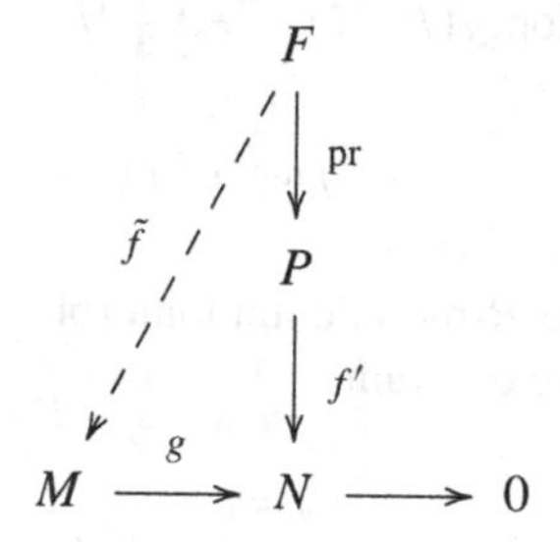

can be completed to a commutative diagram by an oblique arrow $\tilde{f}$. Then

$$g \circ (\tilde{f} \circ i) = (g \circ \tilde{f}) \circ i = f' \circ \mathrm{pr} \circ i = f'$$

which means that the diagram

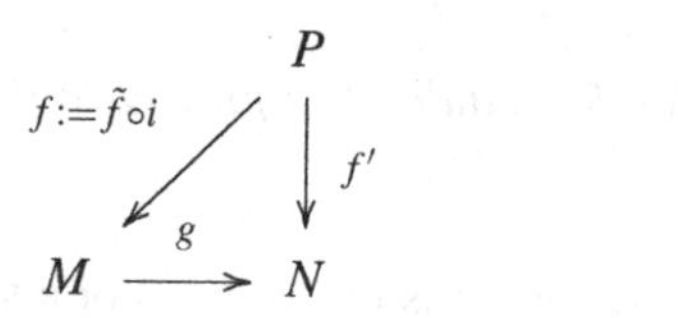

is commutative. Hence P is projective. □

Example Let $e \in R$ be an idempotent. Then $R = Re \oplus R(1 - e)$. Hence Re is a projective R-module.

Corollary 1.6.5 *If P_1, P_2 are two R-modules then the direct sum $P_1 \oplus P_2$ is projective if and only if P_1 and P_2 are projective.*

Proof If

$$(P_1 \oplus P_2) \oplus P' \cong F$$

for some free R-module F then visibly P_1 and P_2 both are isomorphic to direct summands of F as well and hence are projective. If on the other hand

$$P_1 \oplus P_1' \cong \bigoplus_{i\in I_1} R \quad \text{and} \quad P_2 \oplus P_2' \cong \bigoplus_{i\in I_2} R$$

then

$$(P_1 \oplus P_2) \oplus (P_1' \oplus P_2') \cong \bigoplus_{i\in I_1 \cup I_2} R. \qquad \square$$

Lemma 1.6.6 (Schanuel) *Let*

$$0 \longrightarrow L_1 \xrightarrow{h_1} P_1 \xrightarrow{g_1} N \longrightarrow 0 \quad \text{and} \quad 0 \longrightarrow L_2 \xrightarrow{h_2} P_2 \xrightarrow{g_2} N \longrightarrow 0$$

be two short exact sequences of R-modules with the same right-hand term N; if P_1 and P_2 are projective then $L_2 \oplus P_1 \cong L_1 \oplus P_2$.

Proof The R-module

$$M := \{(x_1, x_2) \in P_1 \oplus P_2 : g_1(x_1) = g_2(x_2)\}$$

sets in the two short exact sequences

$$0 \longrightarrow L_2 \xrightarrow{y \mapsto (0, h_2(y))} M \xrightarrow{(x_1, x_2) \mapsto x_1} P_1 \longrightarrow 0$$

and

$$0 \longrightarrow L_1 \xrightarrow{y \mapsto (h_1(y), 0)} M \xrightarrow{(x_1, x_2) \mapsto x_2} P_2 \longrightarrow 0.$$

By applying Lemma 1.6.2 we obtain

$$L_2 \oplus P_1 \cong M \cong L_1 \oplus P_2. \qquad \square$$

Lemma 1.6.7 *Let $R \longrightarrow R'$ be any ring homomorphism; if P is a projective R-module then $R' \otimes_R P$ is a projective R'-module.*

Proof Using Proposition 1.6.4 we write $P \oplus Q \cong \bigoplus_{i\in I} R$ and obtain

$$\begin{aligned}(R' \otimes_R P) \oplus (R' \otimes_R Q) &= R' \otimes_R (P \oplus Q) \cong R' \otimes_R \Big(\bigoplus_{i\in I} R\Big) \\ &= \bigoplus_{i\in I} (R' \otimes_R R) = \bigoplus_{i\in I} R'. \end{aligned} \qquad \square$$

Definition

i. An R-module homomorphism $f : M \longrightarrow N$ is called essential if it is surjective but $f(L) \neq N$ for any proper submodule $L \subsetneq M$.
ii. A projective cover of an R-module M is an essential R-module homomorphism $f : P \longrightarrow M$ where P is projective.

Lemma 1.6.8 *Let $f : P \to M$ and $f' : P' \to M$ be two projective covers; then there exists an R-module isomorphism $g : P' \xrightarrow{\cong} P$ such that $f' = f \circ g$.*

Proof The "test diagram"

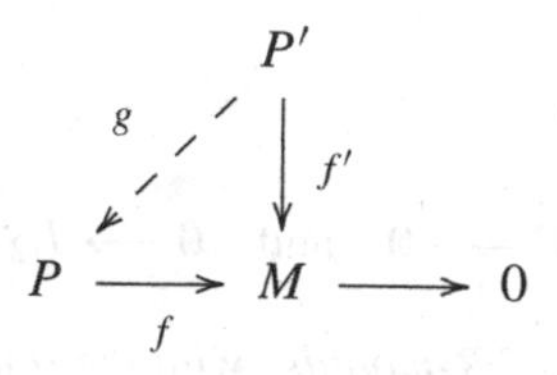

shows the existence of a homomorphism g such that $f' = f \circ g$. Since f' is surjective we have $f(g(P')) = M$, and since f is essential we deduce that $g(P') = P$. This shows that g is surjective. Then, by Lemma 1.6.2, there exists an R-module homomorphism $s : P \longrightarrow P'$ such that $g \circ s = \mathrm{id}_P$. We have

$$f'\big(s(P)\big) = f\big(g\big(s(P)\big)\big) = f(P) = M.$$

Since f' is essential this implies $s(P) = P'$. Hence s and g are isomorphisms. □

Remark 1.6.9 Let $f : M \longrightarrow N$ be a surjective R-module homomorphism between finitely generated R-modules; if

$$\ker(f) \subseteq \mathrm{Jac}(R)M$$

then f is essential.

Proof Let $L \subseteq M$ be a submodule such that $f(L) = N$. Then

$$M = L + \ker(f) = L + \mathrm{Jac}(R)M.$$

Hence $L = M$ by the Nakayama lemma 1.2.3. □

Proposition 1.6.10 *Suppose that R is complete and that $R/\mathrm{Jac}(R)$ is left artinian; then any finitely generated R-module M has a projective cover; more precisely, there exists a projective cover $f : P \longrightarrow M$ such that the induced map $P/\mathrm{Jac}(R)P \xrightarrow{\cong} M/\mathrm{Jac}(R)M$ is an isomorphism.*

Proof By Corollary 1.2.2 and Proposition 1.5.5.i we may write $1+\mathrm{Jac}(R)=\varepsilon_1+\cdots+\varepsilon_r$ as a sum of pairwise orthogonal primitive idempotents $\varepsilon_i \in \overline{R} := R/\mathrm{Jac}(R)$. According to Proposition 1.5.1 and Corollary 1.5.2 we then have

$$\overline{R} = \overline{R}\varepsilon_1 \oplus \cdots \oplus \overline{R}\varepsilon_r$$

where the $\overline{R}$-modules $\overline{R}\varepsilon_i$ are indecomposable. But $\overline{R}$ is semisimple by Proposition 1.2.1.vi. Hence the indecomposable $\overline{R}$-modules $\overline{R}\varepsilon_i$ in fact are simple, and all simple $\overline{R}$-modules occur, up to isomorphism, among the $\overline{R}\varepsilon_i$. On the other hand, $M/\mathrm{Jac}(R)M$ is a semisimple $\overline{R}$-module by Proposition 1.1.4.iii and as such is a direct sum

$$M/\mathrm{Jac}(R)M = \overline{L}_1 \oplus \cdots \oplus \overline{L}_s$$

of simple submodules $\overline{L}_j$. For any $1 \le j \le s$ we find an $1 \le i(j) \le r$ such that

$$\overline{L}_j \cong \overline{R}\varepsilon_{i(j)}.$$

By Proposition 1.5.7 there exist idempotents $e_1, \ldots, e_r \in R$ such that $e_i + \mathrm{Jac}(R) = \varepsilon_i$ for any $1 \le i \le r$. Using Remark 1.5.13 we now consider the composed R-module isomorphism

$$\overline{f}\colon \left(\bigoplus_{j=1}^{s} Re_{i(j)}\right) \Big/ \mathrm{Jac}(R)\left(\bigoplus_{j=1}^{s} Re_{i(j)}\right) \cong \bigoplus_{j=1}^{s} \overline{R}\varepsilon_{i(j)}$$

$$\cong \bigoplus_{j=1}^{s} \overline{L}_j \cong M/\mathrm{Jac}(R)M.$$

It sits in the "test diagram"

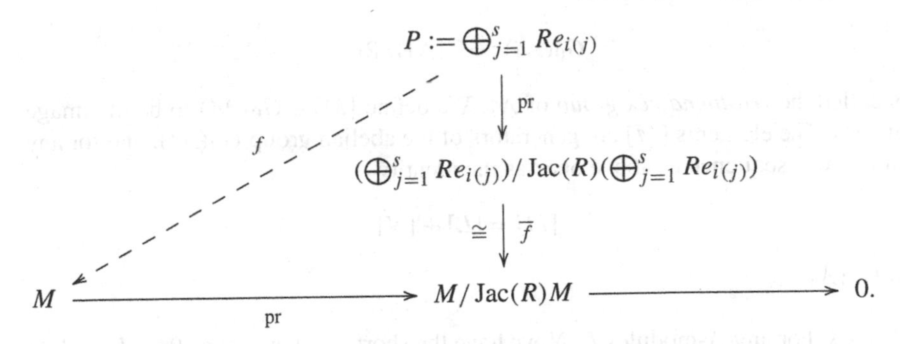

Each Re_i is a projective R-module. Hence P is a projective R-module by Corollary 1.6.5. We therefore find an R-module homomorphism $f : P \longrightarrow M$ which makes the above diagram commutative. By construction it induces the isomorphism $\overline{f}$ modulo $\mathrm{Jac}(R)$. It remains to show that f is essential. Since $\overline{f} \circ \mathrm{pr}$ is surjective we have $f(P) + \mathrm{Jac}(R)M = M$, and the Nakayama lemma 1.2.3 implies that f is surjective. Moreover, $\ker(f) \subseteq \mathrm{Jac}(R)P$ by construction. Hence f is essential by Remark 1.6.9. □

1.7 Grothendieck Groups

We first recall that over any set S one has the free abelian group $\mathbb{Z}[S]$ with basis S given by

$$\mathbb{Z}[S] = \left\{ \sum_{s \in S} m_s s : m_s \in \mathbb{Z}, \text{ all but finitely many } m_s \text{ are equal to zero} \right\}$$

and

$$\left(\sum_{s \in S} m_s s \right) + \left(\sum_{s \in S} n_s s \right) = \sum_{s \in S} (m_s + n_s) s.$$

Its universal property is the following: For any map of sets $\alpha : S \longrightarrow B$ from S into some abelian group B there is a unique homomorphism of abelian groups $\tilde{\alpha} : \mathbb{Z}[S] \longrightarrow B$ such that $\tilde{\alpha}|S = \alpha$.

Let now A be any ring and let $\mathcal{M}$ be some class of A-modules. It is partitioned into isomorphism classes where the isomorphism class $\{M\}$ of a module M in $\mathcal{M}$ consists of all modules in $\mathcal{M}$ which are isomorphic to M. We assume that the isomorphism classes of modules in $\mathcal{M}$ form a set $\mathcal{M}/\cong$, and we introduce the free abelian group $\mathbb{Z}[\mathcal{M}] := \mathbb{Z}[\mathcal{M}/\cong]$. In $\mathbb{Z}[\mathcal{M}]$ we consider the subgroup Rel generated by all elements of the form

$$\{M\} - \{L\} - \{N\}$$

whenever there is a short exact sequence of A-module homomorphisms

$$0 \longrightarrow L \longrightarrow M \longrightarrow N \longrightarrow 0 \quad \text{with } L, M, N \text{ in } \mathcal{M}.$$

The corresponding factor group

$$G_0(\mathcal{M}) := \mathbb{Z}[\mathcal{M}]/\mathrm{Rel}$$

is called the *Grothendieck group* of $\mathcal{M}$. We define $[M] \in G_0(\mathcal{M})$ to be the image of $\{M\}$. The elements $[M]$ are generators of the abelian group $G_0(\mathcal{M})$, and for any short exact sequence as above one has the identity

$$[M] = [L] + [N]$$

in $G_0(\mathcal{M})$.

Remark For any A-modules L, N we have the short exact sequence $0 \to L \to L \oplus N \to N \to 0$ and therefore the identity $[L \oplus N] = [L] + [N]$ in $G_0(\mathcal{M})$ provided L, N, and $L \oplus N$ lie in the class $\mathcal{M}$.

For us two particular cases of this construction will be most important. In the first case we take $\mathfrak{M}_A$ to be the class of all A-modules of finite length, and we define

$$R(A) := G_0(\mathfrak{M}_A).$$

The set $\hat{A}$ of isomorphism classes of simple A-modules is obviously a subset of $\mathfrak{M}_A/\cong$. Hence $\mathbb{Z}[\hat{A}] \subseteq \mathbb{Z}[\mathfrak{M}_A]$ is a subgroup, and we have the composed map

$$\mathbb{Z}[\hat{A}] \xrightarrow{\subseteq} \mathbb{Z}[\mathfrak{M}_A] \xrightarrow{\text{pr}} R(A).$$

Proposition 1.7.1 *The above map* $\mathbb{Z}[\hat{A}] \xrightarrow{\cong} R(A)$ *is an isomorphism.*

Proof We define an endomorphism π of $\mathbb{Z}[\mathfrak{M}_A]$ as follows. By the universal property of $\mathbb{Z}[\mathfrak{M}_A]$ we only need to define $\pi(\{M\}) \in \mathbb{Z}[\mathfrak{M}_A]$ for any A-module M of finite length. Let $\{0\} = M_0 \subsetneqq M_1 \subsetneqq \cdots \subsetneqq M_n = M$ be a composition series of M. According to the Jordan–Hölder Proposition 1.1.2 the isomorphism classes $\{M_1\}, \{M_2/M_1\}, \ldots, \{M/M_{n-1}\}$ do not depend on the choice of the series. We put

$$\pi\big(\{M\}\big) := \{M_1\} + \{M_2/M_1\} + \cdots + \{M/M_{n-1}\},$$

and we observe:

- The modules $M_1, M_2/M_1, \ldots, M/M_{n-1}$ are simple. Hence we have $\text{im}(\pi) \subseteq \mathbb{Z}[\hat{A}]$.
- If M is simple then obviously $\pi(\{M\}) = \{M\}$. It follows that the endomorphism π of $\mathbb{Z}[\mathfrak{M}_A]$ is an idempotent with image equal to $\mathbb{Z}[\hat{A}]$, and therefore

$$\mathbb{Z}[\mathfrak{M}_A] = \text{im}(\pi) \oplus \ker(\pi) = \mathbb{Z}[\hat{A}] \oplus \ker(\pi). \tag{1.7.1}$$

- The exact sequences

$$0 \longrightarrow M_1 \longrightarrow M \longrightarrow M/M_1 \longrightarrow 0$$
$$0 \longrightarrow M_2/M_1 \longrightarrow M/M_1 \longrightarrow M/M_2 \longrightarrow 0$$
$$\vdots$$
$$0 \longrightarrow M_{n-2}/M_{n-1} \longrightarrow M/M_{n-2} \longrightarrow M/M_{n-1} \longrightarrow 0$$

imply the identities

$$[M] = [M_1] + [M/M_1], \qquad [M/M_1] = [M_2/M_1] + [M/M_2], \qquad \ldots,$$
$$[M/M_{n-2}] = [M_{n-1}/M_{n-2}] + [M/M_{n-1}]$$

in $R(A)$. It follows that

$$[M] = [M_1] + [M_2/M_1] + \cdots + [M/M_{n-1}]$$

and therefore $\{M\} - \pi(\{M\}) \in \text{Rel}$. We obtain

$$\mathbb{Z}[\mathfrak{M}_A] = \text{im}(\pi) + \text{Rel} = \mathbb{Z}[\hat{A}] + \text{Rel}. \tag{1.7.2}$$

– Finally, let $0 \longrightarrow L \xrightarrow{f} M \xrightarrow{g} N \longrightarrow 0$ be a short exact sequence with L, M, N in $\mathfrak{M}_A$. Let $\{0\} = L_0 \subsetneqq \ldots \subsetneqq L_r = L$ and $\{0\} = N_0 \subsetneqq \ldots \subsetneqq N_s = N$ be composition series. Then

$$\begin{aligned}\{0\} = M_0 &\subsetneqq M_1 := f(L_1) \subsetneqq \cdots \subsetneqq M_r := f(L) \\ &\subsetneqq M_{r+1} := g^{-1}(N_1) \subsetneqq \cdots \subsetneqq M_{r+s-1} := g^{-1}(N_{s-1}) \subsetneqq M_{r+s} := M\end{aligned}$$

is a composition series of M with

$$M_i/M_{i-1} \cong \begin{cases} L_i/L_{i-1} & \text{for } 1 \le i \le r, \\ N_{i-r}/N_{i-r-1} & \text{for } r < i \le r+s. \end{cases}$$

Hence

$$\begin{aligned}\pi(\{M\}) &= \sum_{i=1}^{r+s} \{M_i/M_{i-1}\} = \sum_{i=1}^{r} \{L_i/L_{i-1}\} + \sum_{j=1}^{s} \{N_j/N_{j-1}\} \\ &= \pi(\{L\}) + \pi(\{N\})\end{aligned}$$

which shows that $\pi(\{M\} - \{L\} - \{N\}) = 0$. It follows that

$$\mathrm{Rel} \subseteq \ker(\pi). \tag{1.7.3}$$

The formulae (1.7.1)–(1.7.3) together imply

$$\mathbb{Z}[\mathfrak{M}_A] = \mathbb{Z}[\hat{A}] \oplus \mathrm{Rel}$$

which is our assertion. □

In the second case we take $\mathcal{M}_A$ to be the class of all finitely generated projective A-modules, and we define

$$K_0(A) := G_0(\mathcal{M}_A).$$

Remark As a consequence of Lemma 1.6.2.ii the subgroup $\mathrm{Rel} \subseteq \mathbb{Z}[\mathcal{M}_A]$ in this case is generated by all elements of the form

$$\{P \oplus Q\} - \{P\} - \{Q\}$$

where P and Q are arbitrary finitely generated projective A-modules.

Let $\tilde{A}$ denote the set of isomorphism classes of finitely generated indecomposable projective A-modules. Then $\mathbb{Z}[\tilde{A}]$ is a subgroup of $\mathbb{Z}[\mathcal{M}_A]$.

Lemma 1.7.2 *If A is left noetherian then the classes $[P]$ for $\{P\} \in \tilde{A}$ are generators of the abelian group $K_0(A)$.*

Proof Let P be any finitely generated projective A-module. By Lemma 1.1.6 we have

$$P = P_1 \oplus \cdots \oplus P_s$$

with finitely generated indecomposable A-modules P_i. The Proposition 1.6.4 implies that the P_i are projective. Hence $\{P_i\} \in \tilde{A}$. As remarked earlier we have

$$[P] = [P_1] + \cdots + [P_s]$$

in $K_0(A)$. □

Remark Under the assumptions of the Krull–Remak–Schmidt Theorem 1.4.7 (e.g., if A is left artinian) an argument completely analogous to the proof of Proposition 1.7.1 shows that the map

$$\begin{aligned} \mathbb{Z}[\tilde{A}] &\xrightarrow{\cong} K_0(A) \\ \{P\} &\longmapsto [P] \end{aligned}$$

is an isomorphism.

But we will see later that, since the unique decomposition into indecomposable modules is needed only for projective modules, weaker assumptions suffice for this isomorphism.

Remark 1.7.3 If A is semisimple we have:

i. $\tilde{A} = \hat{A}$;
ii. any A-module is projective;
iii. $K_0(A) = R(A)$.

Proof By Proposition 1.1.4.iii any A-module is semisimple. In particular, any indecomposable A-module is simple, which proves i. Furthermore, any simple A-module is isomorphic to a module Ae for some idempotent $e \in A$ and hence is projective (compare the proof of Proposition 1.6.10). It follows that any A-module is a direct sum of projective A-modules and therefore is projective (extend the proof of Corollary 1.6.5 to arbitrarily many summands!). This establishes ii. For iii. it remains to note that by Corollary 1.2.2 the class of all finitely generated (projective) A-modules coincides with the class of all A-modules of finite length. □

Suppose that A is left artinian. Then, by Corollary 1.2.2, any finitely generated A-module is of finite length. Hence the homomorphism

$$\begin{aligned} K_0(A) &\longrightarrow R(A) \\ [P] &\longmapsto [P] \end{aligned}$$

is well defined. Using the above remark as well as Proposition 1.7.1 we may introduce the composed homomorphism

$$c_A\colon \quad \mathbb{Z}[\tilde{A}] \xrightarrow{\cong} K_0(A) \longrightarrow R(A) \xrightarrow{\cong} \mathbb{Z}[\hat{A}].$$

It is called the *Cartan homomorphism* of the left artinian ring A. If

$$c_A(\{P\}) = \sum_{\{M\}\in\hat{A}} n_{\{M\}}\{M\}$$

then the integer $n_{\{M\}}$ is the multiplicity with which the simple A-module M occurs, up to isomorphism, as a subquotient in any composition series of the finitely generated indecomposable projective module P.

Let $A \longrightarrow B$ be a ring homomorphism between arbitrary rings A and B. By Lemma 1.6.7 the map

$$\begin{aligned} \mathcal{M}_A/\cong &\longrightarrow \mathcal{M}_B/\cong \\ \{P\} &\longmapsto \{B\otimes_A P\} \end{aligned}$$

and hence the homomorphism

$$\begin{aligned} \mathbb{Z}[\mathcal{M}_A] &\longrightarrow \mathbb{Z}[\mathcal{M}_B] \\ \{P\} &\longmapsto \{B\otimes_A P\} \end{aligned}$$

are well defined. Because of $B\otimes_A(P\oplus Q)\cong (B\otimes_A P)\oplus(B\otimes_A Q)$ the latter map respects the subgroups Rel in both sides. We therefore obtain a well-defined homomorphism

$$\begin{aligned} K_0(A) &\longrightarrow K_0(B) \\ [P] &\longmapsto [B\otimes_A P]. \end{aligned}$$

Exercise Let $I \subseteq A$ be a two-sided ideal. For the projection homomorphism $A \longrightarrow A/I$ and any A-module M we have

$$A/I \otimes_A M = M/IM.$$

In particular, the homomorphism

$$\begin{aligned} K_0(A) &\longrightarrow K_0(A/I) \\ [P] &\longmapsto [P/IP] \end{aligned}$$

is well defined.

Proposition 1.7.4 *Suppose that A is complete and that* $\overline{A} := A/\operatorname{Jac}(A)$ *is left artinian; we then have*:

i. *The maps*

$$\begin{array}{rclcrcl} \tilde{A} & \longrightarrow & \hat{\overline{A}} & \quad \text{and} \quad & K_0(A) & \xrightarrow{\cong} & K_0(\overline{A}) = R(\overline{A}) \\ \{P\} & \longmapsto & \{P/\operatorname{Jac}(A)P\} & & [P] & \longmapsto & [P/\operatorname{Jac}(A)P] \end{array}$$

are bijective;

ii. *the inverses of the maps in i. are given by sending the isomorphism class $\{\overline{M}\}$ of an $\overline{A}$-module $\overline{M}$ of finite length to the isomorphism class of a projective cover of $\overline{M}$ as an A-module;*

iii. $\mathbb{Z}[\tilde{A}] \xrightarrow{\cong} K_0(A)$.

Proof First of all we note that $\overline{A}$ is semisimple by Proposition 1.2.1.iii. We already have seen that the map

$$\begin{aligned} \alpha\colon \quad K_0(A) &\longrightarrow K_0(\overline{A}) = R(\overline{A}) \\ [P] &\longmapsto [P/\operatorname{Jac}(A)P] \end{aligned}$$

is well defined. If $\overline{M}$ is an arbitrary $\overline{A}$-module of finite length then by Proposition 1.6.10 we find a projective cover $P_{\overline{M}} \xrightarrow{f} \overline{M}$ of $\overline{M}$ as an A-module such that

$$P_{\overline{M}}/\operatorname{Jac}(A)P_{\overline{M}} \xrightarrow{\cong} \overline{M}. \tag{1.7.4}$$

The proof of Proposition 1.6.10 shows that $P_{\overline{M}}$ in fact is a finitely generated A-module. According to Lemma 1.6.8 the isomorphism class $\{P_{\overline{M}}\}$ only depends on the isomorphism class $\{\overline{M}\}$. We conclude that

$$\begin{aligned} \mathbb{Z}[\mathcal{M}_{\overline{A}}] &\longrightarrow \mathbb{Z}[\mathcal{M}_A] \\ \{\overline{M}\} &\longmapsto \{P_{\overline{M}}\} \end{aligned}$$

is a well-defined homomorphism. If $\overline{N}$ is a second $\overline{A}$-module of finite length with projective cover $P_{\overline{N}} \xrightarrow{g} \overline{N}$ as above then

$$P_{\overline{M}} \oplus P_{\overline{N}} \xrightarrow{f\oplus g} \overline{M} \oplus \overline{N}$$

is surjective with

$$\begin{aligned} (P_{\overline{M}} \oplus P_{\overline{N}})/\operatorname{Jac}(A)(P_{\overline{M}} \oplus P_{\overline{N}}) &= P_{\overline{M}}/\operatorname{Jac}(A)P_{\overline{M}} \oplus P_{\overline{N}}/\operatorname{Jac}(A)P_{\overline{N}} \\ &\cong \overline{M} \oplus \overline{N}. \end{aligned}$$

It follows that $\ker(f \oplus g) = \operatorname{Jac}(A)(P_{\overline{M}} \oplus P_{\overline{N}})$. This, by Remark 1.6.9, implies that $f \oplus g$ is essential. Using Corollary 1.6.5 we see that $f \oplus g$ is a projective cover of $\overline{M} \oplus \overline{N}$ as an A-module. Hence we have

$$\{P_{\overline{M}} \oplus P_{\overline{N}}\} = \{P_{\overline{M}\oplus\overline{N}}\}.$$

This means that the above map respects the subgroups Rel in both sides and consequently induces a homomorphism

$$\begin{aligned} \beta \colon \quad K_0(\overline{A}) &\longrightarrow K_0(A) \\ [\overline{M}] &\longmapsto [P_{\overline{M}}]. \end{aligned}$$

But it also shows that $\overline{M}$ is a simple $\overline{A}$-module if $P_{\overline{M}}$ is an indecomposable A-module. The isomorphism (1.7.4) says that

$$\alpha \circ \beta = \mathrm{id}_{K_0(\overline{A})} .$$

On the other hand, let P be any finitely generated projective A-module. As a consequence of Remark 1.6.9 the projection map $P \longrightarrow \overline{M} := P/\operatorname{Jac}(A)P$ is essential and hence a projective cover. We then deduce from Lemma 1.6.8 that $\{P\} = \{P_{\overline{M}}\}$ which means that

$$\beta \circ \alpha = \mathrm{id}_{K_0(A)} .$$

It follows that α is an isomorphism with inverse β. We also see that if $P = P_1 \oplus P_2$ is decomposable then $\overline{M} = P_1/\operatorname{Jac}(A)P_1 \oplus P_2/\operatorname{Jac}(A)P_2$ is decomposable as well. This establishes the assertions i. and ii. For iii. we consider the commutative diagram

$$\begin{array}{ccc} K_0(A) & \xrightarrow{\cong} & R(\overline{A}) \\ \uparrow & & \uparrow \cong \\ \mathbb{Z}[\tilde{A}] & \xrightarrow{\cong} & \mathbb{Z}[\hat{\overline{A}}] \end{array}$$

where the horizontal isomorphisms come from i. and the right vertical isomorphism was shown in Proposition 1.7.1. Hence the left vertical arrow is an isomorphism as well. □

Corollary 1.7.5 *Suppose that A is complete and that $A/\operatorname{Jac}(A)$ is left artinian; then*

$$A = P_1 \oplus \cdots \oplus P_r$$

decomposes into a direct sum of finitely many finitely generated indecomposable projective A-modules P_j, and any finitely generated indecomposable projective A-module is isomorphic to one of the P_j.

Proof The projection map $A \longrightarrow A/\operatorname{Jac}(A)$ is a projective cover. We now decompose the semisimple ring

$$A/\operatorname{Jac}(A) = \overline{M_1} \oplus \cdots \oplus \overline{M_r}$$

as a direct sum of finitely many simple modules $\overline{M_j}$, and we choose projective covers $P_j \longrightarrow \overline{M_j}$. Then $P_1 \oplus \cdots \oplus P_r$ is a projective cover of $A/\operatorname{Jac}(A)$ and consequently is isomorphic to A. If P is an arbitrary finitely generated indecomposable projective A-module then P is a projective cover of the simple module $P/\operatorname{Jac}(A)P$. The latter has to be isomorphic to some $\overline{M_j}$. Hence $P \cong P_j$. □

Assuming that A is left artinian let us go back to the Cartan homomorphism

$$c_A\colon \quad \mathbb{Z}[\hat{A}] \cong K_0(A) \longrightarrow R(A) \cong \mathbb{Z}[\hat{A}].$$

By Corollary 1.7.5 the set $\tilde{A} = \{\{P_1\}, \ldots, \{P_t\}\}$ is finite. We put

$$M_j := P_j / \operatorname{Jac}(A) P_j.$$

Then, by Proposition 1.7.4.i, $\{M_1\}, \ldots, \{M_t\}$ are exactly the isomorphism classes of simple $A/\operatorname{Jac}(A)$-modules. But due to the definition of the Jacobson radical the simple $A/\operatorname{Jac}(A)$-modules coincide with the simple A-modules, i.e.

$$\hat{A} = \big\{\{M_1\}, \ldots, \{M_t\}\big\}.$$

The Cartan homomorphism therefore is given by an integral matrix

$$C_A = (c_{ij})_{1 \le i,j \le t} \in M_{t \times t}(\mathbb{Z})$$

defined by the equations

$$c_A\big(\{P_j\}\big) = c_{1j}\{M_1\} + \cdots + c_{tj}\{M_t\}.$$

The matrix C_A is called the *Cartan matrix* of A.

Chapter 2
The Cartan–Brauer Triangle

Let G be a finite group. Over any commutative ring R we have the group ring

$$R[G] = \left\{ \sum_{g \in G} a_g g : a_g \in R \right\}$$

with addition

$$\left(\sum_{g \in G} a_g g \right) + \left(\sum_{g \in G} b_g g \right) = \sum_{g \in G} (a_g + b_g) g$$

and multiplication

$$\left(\sum_{g \in G} a_g g \right) \left(\sum_{g \in G} b_g g \right) = \sum_{g \in G} \left(\sum_{h \in G} a_h b_{h^{-1}g} \right) g.$$

We fix an algebraically closed field k of characteristic $p > 0$. The representation theory of G over k is the module theory of the group ring $k[G]$. This is our primary object of study in the following.

2.1 The Setting

The main technical tool of our investigation will be a $(0, p)$*-ring* R for k which is a complete local commutative integral domain R such that

- the maximal ideal $\mathfrak{m}_R \subseteq R$ is principal,
- $R/\mathfrak{m}_R = k$, and
- the field of fractions of R has characteristic zero.

Exercise The only ideals of R are $\mathfrak{m}_R^j$ for $j \geq 0$ and $\{0\}$.

P. Schneider, *Modular Representation Theory of Finite Groups*,
DOI 10.1007/978-1-4471-4832-6_2, © Springer-Verlag London 2013

We note that there must exist an integer $e \geq 1$—the *ramification index* of R—such that $Rp = \mathfrak{m}_R^e$. There is, in fact, a canonical $(0, p)$-ring $W(k)$ for k—its *ring of Witt vectors*—with the additional property that $\mathfrak{m}_{W(k)} = W(k)p$. Let K/K_0 be any finite extension of the field of fractions K_0 of $W(k)$. Then

$$R := \{a \in K : \mathrm{Norm}_{K/K_0}(a) \in W(k)\}$$

is a $(0, p)$-ring for k with ramification index equal to $[K : K_0]$. Proofs for all of this can be found in §3–6 of [9].

In the following we fix a $(0, p)$-ring R for k. We denote by K the field of fractions of R and by π_R a choice of generator of $\mathfrak{m}_R$, i.e. $\mathfrak{m}_R = R\pi_R$. The following three group rings are now at our disposal:

$$\begin{array}{ccc} K[G] & & \\ \subseteq \uparrow & & \\ R[G] & \xrightarrow{\mathrm{pr}} & k[G] \end{array}$$

such that

$$K[G] = K \otimes_R R[G] \quad \text{and} \quad k[G] = R[G]/\pi_R R[G].$$

As explained before Proposition 1.7.4 there are the corresponding homomorphisms between Grothendieck groups

$$\begin{array}{ccc} K_0(K[G]) & & \\ {\scriptstyle [P] \mapsto [K \otimes_R P]} \uparrow {\scriptstyle \kappa} & & \\ K_0(R[G]) & \xrightarrow[{[P] \mapsto [P/\pi_R P]}]{\rho} & K_0(k[G]). \end{array}$$

For the vertical arrow observe that, quite generally for any $R[G]$-module M, we have

$$K[G] \otimes_{R[G]} M = K \otimes_R R[G] \otimes_{R[G]} M = K \otimes_R M.$$

We put

$$R_K(G) := R(K[G]) \quad \text{and} \quad R_k(G) := R(k[G]).$$

Since K has characteristic zero the group ring $K[G]$ is semisimple, and we have

$$R_K(G) = K_0(K[G])$$

by Remark 1.7.3.iii. On the other hand, as a finite-dimensional k-vector space the group ring $k[G]$ of course is left and right artinian. In particular we have the Cartan homomorphism

$$c_G\colon\ K_0\big(k[G]\big) \longrightarrow R_k(G)$$
$$[P] \longmapsto [P].$$

Hence, so far, there is the diagram of homomorphisms

$$\begin{array}{ccc} R_K(G) & & R_k(G) \\ \uparrow{\scriptstyle\kappa} & & \uparrow{\scriptstyle c_G} \\ K_0(R[G]) & \xrightarrow{\ \rho\ } & K_0(k[G]). \end{array}$$

Clearly, $R[G]$ is an R-algebra which is finitely generated as an R-module. Let us collect some of what we know in this situation.

- (Proposition 1.3.6) $R[G]$ is left and right noetherian, and any finitely generated $R[G]$-module is complete as well as $R[G]\pi_R$-adically complete.
- (Theorem 1.4.7) The Krull–Remak–Schmidt theorem holds for any finitely generated $R[G]$-module.
- (Proposition 1.5.5) $1 \in R[G]$ can be written as a sum of pairwise orthogonal primitive idempotents; the set of all central idempotents in $R[G]$ is finite; 1 is equal to the sum of all primitive idempotents in $Z(R[G])$; any $R[G]$-module has a block decomposition.
- (Proposition 1.5.7) For any idempotent $\varepsilon \in k[G]$ there is an idempotent $e \in R[G]$ such that $\varepsilon = e + R[G]\pi_R$.
- (Proposition 1.5.12) The projection map $R[G] \longrightarrow k[G]$ restricts to a bijection between the set of all central idempotents in $R[G]$ and the set of all central idempotents in $k[G]$; in particular, the block decomposition of an $R[G]$-module M reduces modulo $R[G]\pi_R$ to the block decomposition of the $k[G]$-module $M/\pi_R M$.
- (Proposition 1.6.10) Any finitely generated $R[G]$-module M has a projective cover $P \longrightarrow M$ such that $P/\operatorname{Jac}(R[G])P \xrightarrow{\cong} M/\operatorname{Jac}(R[G])M$ is an isomorphism.

We emphasize that $R[G]\pi_R \subseteq \operatorname{Jac}(R[G])$ by Lemma 1.3.5.iii. Moreover, the ideal $\operatorname{Jac}(k[G]) = \operatorname{Jac}(R[G])/R[G]\pi_R$ in the left artinian ring $k[G]$ is nilpotent by Proposition 1.2.1.vi. We apply Proposition 1.7.4 to the rings $R[G]$ and $k[G]$ and we see that the maps $\{P\} \longmapsto \{P/\pi_R P\} \longmapsto \{P/\operatorname{Jac}(R[G])P\}$ induce the commutative diagram of bijections between finite sets

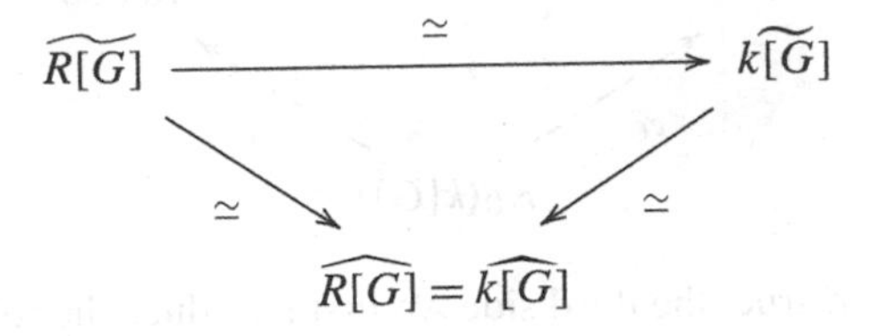

as well as the commutative diagram of isomorphisms

$$\begin{array}{ccc} \mathbb{Z}[\widetilde{R[G]}] & \xrightarrow{\cong} & \mathbb{Z}[\widetilde{k[G]}] \\ \cong\downarrow & & \downarrow\cong \\ K_0(R[G]) & \underset{\rho}{\xrightarrow{\cong}} & K_0(k[G]). \end{array}$$

For purposes of reference we state the last fact as a proposition.

Proposition 2.1.1 *The map $\rho : K_0(R[G]) \xrightarrow{\cong} K_0(k[G])$ is an isomorphism; its inverse is given by sending $[M]$ to the class of a projective cover of M as an $R[G]$-module.*

We define the composed map

$$e_G: \quad K_0\big(k[G]\big) \xrightarrow{\rho^{-1}} K_0\big(R[G]\big) \xrightarrow{\kappa} K_0\big(K[G]\big) = R_K(G).$$

Remark 2.1.2 Any finitely generated projective R-module is free.

Proof Since R is an integral domain 1 is the only idempotent in R. Hence the free R-module R is indecomposable by Corollary 1.5.2. On the other hand, according to Proposition 1.7.4.i the map

$$\begin{aligned} \tilde{R} &\xrightarrow{\simeq} \hat{k} \\ \{P\} &\longmapsto \{P/\pi_R P\} \end{aligned}$$

is bijective. Obviously, k is up to isomorphism the only simple k-module. Hence R is up to isomorphism the only finitely generated indecomposable projective R-module. An arbitrary finitely generated projective R-module P, by Lemma 1.1.6, is a finite direct sum of indecomposable ones. It follows that P must be isomorphic to some R^n. □

2.2 The Triangle

We already have the two sides

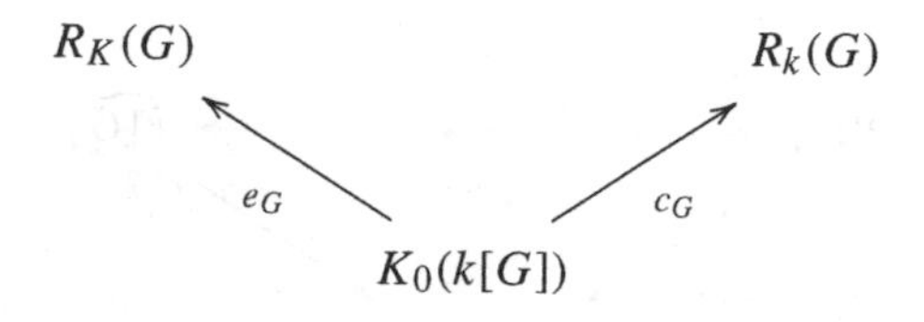

of the triangle. To construct the third side we first introduce the following notion.

Definition Let V be a finite-dimensional K-vector space; a lattice L in V is an R-submodule $L \subseteq V$ for which there exists a K-basis $e_1, \ldots, e_d$ of V such that

$$L = Re_1 + \cdots + Re_d.$$

Obviously, any lattice is free as an R-module. Furthermore, with L also aL, for any $a \in K^\times$, is a lattice in V.

Lemma 2.2.1

i. *Let L be an R-submodule of a K-vector space V; if L is finitely generated then L is free.*
ii. *Let $L \subseteq V$ be an R-submodule of a finite-dimensional K-vector space V; if L is finitely generated as an R-module and L generates V as a K-vector space then L is a lattice in V.*
iii. *For any two lattices L and L' in V there is an integer $m \geq 0$ such that $\pi_R^m L \subseteq L'$.*

Proof i. and ii. Let $d \geq 0$ be the smallest integer such that the R-module L has d generators $e_1, \ldots, e_d$. The R-module homomorphism

$$R^d \longrightarrow L$$
$$(a_1, \ldots, a_d) \longmapsto a_1 e_1 + \cdots + a_d e_d$$

is surjective. Suppose that $(a_1, \ldots, a_d) \neq 0$ is an element in its kernel. Since at least one a_i is nonzero the integer

$$\ell := \max\{j \geq 0 : a_1, \ldots, a_d \in \mathfrak{m}_R^j\}$$

is defined. Then $a_i = \pi_R^\ell b_i$ with $b_i \in R$, and $b_{i_0} \in R^\times$ for at least one index $1 \leq i_0 \leq d$. Computing in the vector space V we have

$$0 = a_1 e_1 + \cdots + a_d e_d = \pi_R^\ell (b_1 e_1 + \cdots + b_d e_d)$$

and hence

$$b_1 e_1 + \cdots + b_d e_d = 0.$$

But the latter equation implies $e_{i_0} = -\sum_{i \neq i_0} b_{i_0}^{-1} b_i e_i \in \sum_{i \neq i_0} Re_i$, which is a contradiction to the minimality of d. It follows that the above map is an isomorphism. This proves i. and, in particular, that

$$L = Re_1 + \cdots + Re_d.$$

For ii. it therefore suffices to show that $e_1, \ldots, e_d$, under the additional assumption that L generates V, is a K-basis of V. This assumption immediately guarantees that the $e_1, \dots, e_d$ generate the K-vector space V. To show that they are K-linearly independent let

$$c_1e_1 + \cdots + c_de_d = 0 \quad \text{with } c_1, \dots, c_d \in K.$$

We find a sufficiently large $j \geq 0$ such that $a_i := \pi_R^j c_i \in R$ for any $1 \leq i \leq d$. Then

$$0 = \pi_R^\ell \cdot 0 = \pi_R^\ell (c_1e_1 + \cdots + c_de_d) = a_1e_1 + \cdots + a_de_d.$$

By what we have shown above we must have $a_i = 0$ and hence $c_i = 0$ for any $1 \leq i \leq d$.

iii. Let $e_1, \dots, e_d$ and $f_1, \dots, f_d$ be K-bases of V such that

$$L = Re_1 + \cdots + Re_d \quad \text{and} \quad L' = Rf_1 + \cdots + Rf_d.$$

We write

$$e_j = c_{1j}f_1 + \cdots + c_{dj}f_d \quad \text{with } c_{ij} \in K,$$

and we choose an integer $m \geq 0$ such that

$$\pi_R^m c_{ij} \in R \quad \text{for any } 1 \leq i, j \leq d.$$

It follows that $\pi_R^m e_j \in Rf_1 + \cdots + Rf_d = L'$ for any $1 \leq j \leq d$ and hence that $\pi_R^m L \subseteq L'$. □

Suppose that V is a finitely generated $K[G]$-module. Then V is finite-dimensional as a K-vector space. A lattice L in V is called *G-invariant* if we have $gL \subseteq L$ for any $g \in G$. In particular, L is a finitely generated $R[G]$-submodule of V, and $L/\pi_R L$ is a $k[G]$-module of finite length.

Lemma 2.2.2 *Any finitely generated $K[G]$-module V contains a lattice which is G-invariant.*

Proof We choose a basis $e_1, \dots, e_d$ of the K-vector space V. Then $L' := Re_1 + \cdots + Re_d$ is a lattice in V. We define the $R[G]$-submodule

$$L := \sum_{g \in G} gL'$$

of V. With L' also L generates V as a K-vector space. Moreover, L is finitely generated by the set $\{ge_i : 1 \leq i \leq d, g \in G\}$ as an R-module. Therefore, by Lemma 2.2.1.ii, L is a G-invariant lattice in V. □

Whereas the $K[G]$-module V always is projective by Remark 1.7.3 a G-invariant lattice L in V need not to be projective as an $R[G]$-module. We will encounter an example of this later on.

Theorem 2.2.3 *Let L and L' be two G-invariant lattices in the finitely generated $K[G]$-module V; we then have*

$$[L/\pi_R L] = [L'/\pi_R L'] \quad \text{in } R_k(G).$$

Proof We begin by observing that, for any $a \in K^\times$, the map

$$L/\pi_R L \xrightarrow{\cong} (aL)/\pi_R(aL)$$
$$x + \pi_R L \longmapsto ax + \pi_R(aL)$$

is an isomorphism of $k[G]$-modules, and hence

$$[L/\pi_R L] = [(aL)/\pi_R(aL)] \quad \text{in } R_k(G).$$

By applying Lemma 2.2.1.iii (and replacing L by $\pi_R^m L$ for some sufficiently large $m \geq 0$) we therefore may assume that $L \subseteq L'$. By applying Lemma 2.2.1.iii again to L and L' (while interchanging their roles) we find an integer $n \geq 0$ such that

$$\pi_R^n L' \subseteq L \subseteq L'.$$

We now proceed by induction with respect to n. If $n = 1$ we have the two exact sequences of $k[G]$-modules

$$0 \longrightarrow L/\pi_R L' \longrightarrow L'/\pi_R L' \longrightarrow L'/L \longrightarrow 0$$

and

$$0 \longrightarrow \pi_R L'/\pi_R L \longrightarrow L/\pi_R L \longrightarrow L/\pi_R L' \longrightarrow 0.$$

It follows that

$$\begin{aligned} [L'/\pi_R L'] &= [L/\pi_R L'] + [L'/L] = [L/\pi_R L'] + [\pi_R L'/\pi_R L] \\ &= [L/\pi_R L'] + [L/\pi_R L] - [L/\pi_R L'] \\ &= [L/\pi_R L] \end{aligned}$$

in $R_k(G)$. For $n \geq 2$ we consider the $R[G]$-submodule

$$M := \pi_R^{n-1} L' + L.$$

It is a G-invariant lattice in V by Lemma 2.2.1.ii and satisfies

$$\pi_R^{n-1} L' \subseteq M \subseteq L' \quad \text{and} \quad \pi_R M \subseteq L \subseteq M.$$

Applying the case $n = 1$ to L and M we obtain $[M/\pi_R M] = [L/\pi_R L]$. The induction hypothesis for M and L' gives $[L'/\pi_R L'] = [M/\pi_R M]$. □

The above lemma and theorem imply that

$$\begin{aligned} \mathbb{Z}[\mathcal{M}_{K[G]}] &\longrightarrow R_k(G) \\ \{V\} &\longmapsto [L/\pi_R L], \end{aligned}$$

where L is any G-invariant lattice in V, is a well-defined homomorphism. If V_1 and V_2 are two finitely generated $K[G]$-modules and $L_1 \subseteq V_1$ and $L_2 \subseteq V_2$ are G-invariant lattices then $L_1 \oplus L_2$ is a G-invariant lattice in $V_1 \oplus V_2$ and

$$\begin{aligned} \big[(L_1 \oplus L_2)/\pi_R(L_1 \oplus L_2)\big] &= [L_1/\pi_R L_1 \oplus L_2/\pi_R L_2] \\ &= [L_1/\pi_R L_1] + [L_2/\pi_R L_2]. \end{aligned}$$

It follows that the subgroup $\mathrm{Rel} \subseteq \mathbb{Z}[\mathcal{M}_{K[G]}]$ lies in the kernel of the above map so that we obtain the homomorphism

$$\begin{aligned} d_G\colon \quad R_K(G) &\longrightarrow R_k(G) \\ [V] &\longmapsto [L/\pi_R L]. \end{aligned}$$

It is called the *decomposition homomorphism* of G. The *Cartan–Brauer triangle* is the diagram

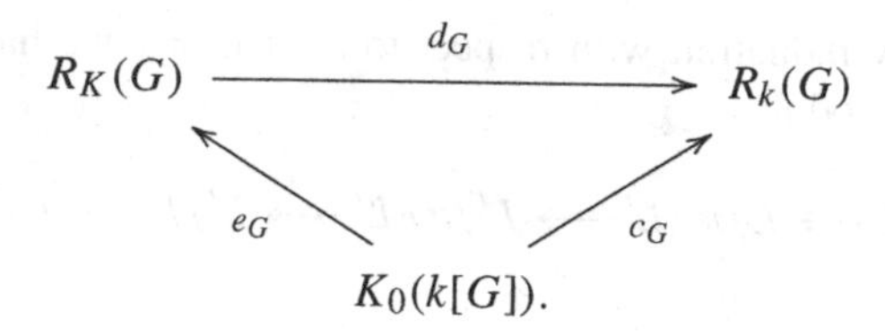

Lemma 2.2.4 *The Cartan–Brauer triangle is commutative.*

Proof Let P be a finitely generated projective $R[G]$-module. We have to show that

$$d_G\big(\kappa([P])\big) = c_G\big(\rho([P])\big)$$

holds true. By definition the right-hand side is equal to $[P/\pi_R P] \in R_k(G)$. Moreover, $\kappa([P]) = [K \otimes_R P] \in R_K(G)$. According to Proposition 1.6.4 the $R(G)$-module P is a direct summand of a free $R[G]$-module. But $R[G]$ and hence any free $R[G]$-module also is free as an R-module. We see that P as an R-module is finitely generated projective and hence free by Remark 2.1.2. We conclude that $P \cong R^d$ is a G-invariant lattice in the $K[G]$-module $K \otimes_R P \cong K \otimes_R R^d = (K \otimes_R R)^d = K^d$, and we obtain $d_G(\kappa([P])) = d_G([K \otimes_R P]) = [P/\pi_R P]$. □

Let us consider two "extreme" situations where the maps in the Cartan–Brauer triangle can be determined completely. First we look at the case where p does not divide the order $|G|$ of the group G. Then $k[G]$ is semisimple. Hence we have $\widetilde{k[G]} = \widehat{k[G]}$ and $K_0(k[G]) = R_k(G)$ by Remark 1.7.3. The map c_G, in particular, is the identity.

Proposition 2.2.5 *If $p \nmid |G|$ then any $R[G]$-module M which is projective as an R-module also is projective as an $R[G]$-module.*

Proof We consider any "test diagram" of $R[G]$-modules

$$\begin{array}{ccccc} & & M & & \\ & & \downarrow{\scriptstyle \alpha} & & \\ L & \xrightarrow{\beta} & N & \longrightarrow & 0. \end{array}$$

Viewing this as a "test diagram" of R-modules our second assumption ensures the existence of an R-module homomorphism $\alpha_0 : M \longrightarrow L$ such that $\beta \circ \alpha_0 = \alpha$. Since $|G|$ is a unit in R by our first assumption, we may define a new R-module homomorphism $\tilde{\alpha} : M \longrightarrow L$ by

$$\tilde{\alpha}(x) := |G|^{-1} \sum_{g \in G} g\alpha_0(g^{-1}x) \quad \text{for any } x \in M.$$

One easily checks that $\tilde{\alpha}$ satisfies

$$\tilde{\alpha}(hx) = h\tilde{\alpha}(x) \quad \text{for any } h \in G \text{ and any } x \in M.$$

This means that $\tilde{\alpha}$ is, in fact, an $R[G]$-module homomorphism. Moreover, we compute

$$\begin{aligned} \beta(\tilde{\alpha}(x)) &= |G|^{-1} \sum_{g \in G} g\beta(\alpha_0(g^{-1}x)) = |G|^{-1} \sum_{g \in G} g\alpha(g^{-1}x) \\ &= |G|^{-1} \sum_{g \in G} gg^{-1}\alpha(x) = \alpha(x). \end{aligned}$$

□

Corollary 2.2.6 *If $p \nmid |G|$ then all three maps in the Cartan–Brauer triangle are isomorphisms; more precisely, we have the triangle of bijections*

$$\begin{array}{ccc} \widehat{K[G]} & \xrightarrow{\simeq} & \widehat{k[G]} \\ {\scriptstyle \{P\} \mapsto \{K \otimes_R P\}} \nwarrow \simeq & & \simeq \nearrow {\scriptstyle \{P\} \mapsto \{P/\pi_R P\}} \\ & \widetilde{R[G]}. & \end{array}$$

Proof We already have remarked that c_G is the identity. Hence it suffices to show that the map $\kappa : K_0(R[G]) \longrightarrow R_K(G)$ is surjective. Let V be any finitely generated $K[G]$-module. By Lemma 2.2.2 we find a G-invariant lattice L in V. It satisfies $V = K \otimes_R L$ by definition. Proposition 2.2.5 implies that L is a finitely generated projective $R[G]$-module. We conclude that $[L] \in K_0(R[G])$ with $\kappa([L]) = [V]$. This argument in fact shows that the map

$$\begin{aligned} \mathcal{M}_{R[G]}/\cong &\longrightarrow \mathcal{M}_{K[G]}/\cong \\ \{P\} &\longmapsto \{K \otimes_R P\} \end{aligned}$$

is surjective. Let P and Q be two finitely generated projective $R[G]$-modules such that $K \otimes_R P \cong K \otimes_R Q$ as $K[G]$-modules. The commutativity of the Cartan–Brauer triangle then implies that

$$[P/\pi_R P] = d_G\big([K \otimes_R P]\big) = d_G\big([K \otimes_R Q]\big) = [Q/\pi_R Q].$$

Let $P = P_1 \oplus \cdots \oplus P_s$ and $Q = Q_1 \oplus \cdots \oplus Q_t$ be decompositions into indecomposable submodules. Then

$$P/\pi_R P = \bigoplus_{i=1}^{s} P_i/\pi_R P_i \quad \text{and} \quad Q/\pi_R Q = \bigoplus_{j=1}^{t} Q_j/\pi_R Q_j$$

are decompositions into simple submodules. Using Proposition 1.7.1 the identity

$$\sum_{i=1}^{s} [P_i/\pi_R P_i] = [P/\pi_R P] = [Q/\pi_R Q] = \sum_{j=1}^{t} [Q_j/\pi_R Q_j]$$

implies that $s = t$ and that there is a permutation σ of $\{1, \ldots, s\}$ such that

$$Q_j/\pi_R Q_j \cong P_{\sigma(j)}/\pi_R P_{\sigma(j)} \quad \text{for any } 1 \le j \le s$$

as $k[G]$-modules. Applying Proposition 1.7.4.i we obtain

$$Q_j \cong P_{\sigma(j)} \quad \text{for any } 1 \le j \le s$$

as $R[G]$-modules. It follows that $P \cong Q$ as $R[G]$-modules. Hence the above map between sets of isomorphism classes is bijective. Obviously, if $K \otimes_R P$ is indecomposable (i.e. simple) then P was indecomposable. Vice versa, if P is indecomposable then the above reasoning says that $P/\pi_R P$ is simple. Because of $[P/\pi_R P] = d_G([K \otimes_R P])$ it follows that $K \otimes_R P$ must be indecomposable. □

The second case is where G is a p-group. For a general group G we have the ring homomorphism

$$\begin{aligned} k[G] &\longrightarrow k \\ \sum_{g \in G} a_g g &\longmapsto \sum_{g \in G} a_g \end{aligned}$$

which is called the *augmentation* of $k[G]$. It makes k into a simple $k[G]$-module which is called the *trivial $k[G]$-module*. Its kernel is the *augmentation ideal*

$$I_k[G] := \left\{ \sum_{g \in G} a_g g \in k[G] : \sum_{g \in G} a_g = 0 \right\}.$$

Proposition 2.2.7 *If G is a p-group then we have* $\mathrm{Jac}(k[G]) = I_k[G]$; *in particular, $k[G]$ is a local ring and the trivial $k[G]$-module is, up to isomorphism, the only simple $k[G]$-module.*

Proof We will prove by induction with respect to the order $|G| = p^n$ of G that the trivial module, up to isomorphism, is the only simple $k[G]$-module. There is nothing to prove if $n = 0$. We therefore suppose that $n \geq 1$. Note that we have

$$g^{p^n} = 1 \quad \text{and hence} \quad (g-1)^{p^n} = g^{p^n} - 1 = 1 - 1 = 0$$

for any $g \in G$. The center of a nontrivial p-group is nontrivial. Let $g_0 \neq 1$ be a central element in G. We now consider any simple $k[G]$-module M and we denote by $\pi : k[G] \longrightarrow \mathrm{End}_k(M)$ the corresponding ring homomorphism. Then

$$\left(\pi(g_0) - \mathrm{id}_M\right)^{p^n} = \pi(g_0 - 1)^{p^n} = \pi\left((g_0 - 1)^{p^n}\right) = \pi(0) = 0.$$

Since g_0 is central $(\pi(g_0) - \mathrm{id}_M)(M)$ is a $k[G]$-submodule of M. But M is simple. Hence $(\pi(g_0) - \mathrm{id}_M)(M) = \{0\}$ or $= M$. The latter would inductively imply that $(\pi(g_0) - \mathrm{id}_M)^{p^n}(M) = M \neq \{0\}$ which is contradiction. We obtain $\pi(g_0) = \mathrm{id}_M$ which means that the cyclic subgroup $\langle g_0 \rangle$ is contained in the kernel of the group homomorphism $\pi : G \longrightarrow \mathrm{Aut}_k(M)$. Hence we have a commutative diagram of group homomorphisms

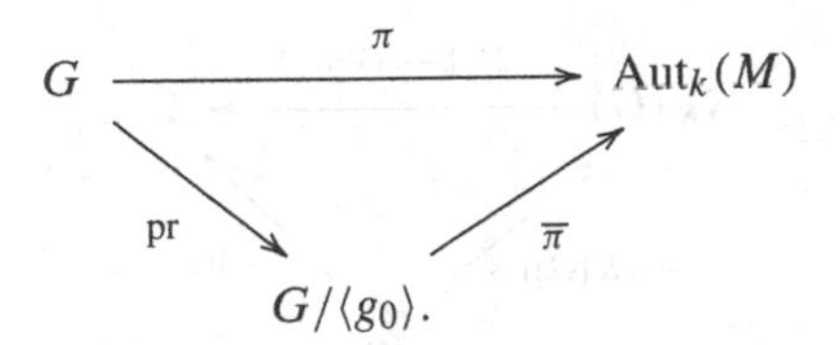

We conclude that M already is a simple $k[G/\langle g_0 \rangle]$-module and therefore is the trivial module by the induction hypothesis.

The identity $\mathrm{Jac}(k[G]) = I_k[G]$ now follows from the definition of the Jacobson radical, and Proposition 1.4.1 implies that $k[G]$ is a local ring. □

Suppose that G is a p-group. Then Propositions 2.2.7 and 1.7.1 together imply that the map

$$\begin{aligned} \mathbb{Z} &\xrightarrow{\cong} R_k(G) \\ m &\longmapsto m[k] \end{aligned}$$

is an isomorphism. To compute the inverse map let M be a $k[G]$-module of finite length, and let $\{0\} = M_0 \subsetneqq M_1 \subsetneqq \cdots \subsetneqq M_t = M$ be a composition series. We must have $M_i/M_{i-1} \cong k$ for any $1 \le i \le t$. It follows that

$$[M] = \sum_{i=1}^{t} [M_i/M_{i-1}] = t \cdot [k] \quad \text{and} \quad \dim_k M = \sum_{i=1}^{t} \dim_k M_i/M_{i-1} = t.$$

Hence the inverse map is given by

$$[M] \longmapsto \dim_k M.$$

Furthermore, from Proposition 1.7.4 we obtain that

$$\begin{aligned} \mathbb{Z} &\xrightarrow{\cong} K_0\bigl(R[G]\bigr) \\ m &\longmapsto m\bigl[R[G]\bigr] \end{aligned}$$

is an isomorphism. Because of $\dim_k k[G] = |G|$ the Cartan homomorphism, under these identifications, becomes the map

$$\begin{aligned} c_G\colon\ \mathbb{Z} &\longrightarrow \mathbb{Z} \\ m &\longmapsto m \cdot |G|. \end{aligned}$$

For any finitely generated $K[G]$-module V and any G-invariant lattice L in V we have $\dim_K V = \dim_k L/\pi_R L$. Hence the decomposition homomorphism becomes

$$\begin{aligned} d_G\colon\ R_K(G) &\longrightarrow \mathbb{Z} \\ [V] &\longmapsto \dim_K V. \end{aligned}$$

Altogether the Cartan–Brauer triangle of a p-group G is of the form

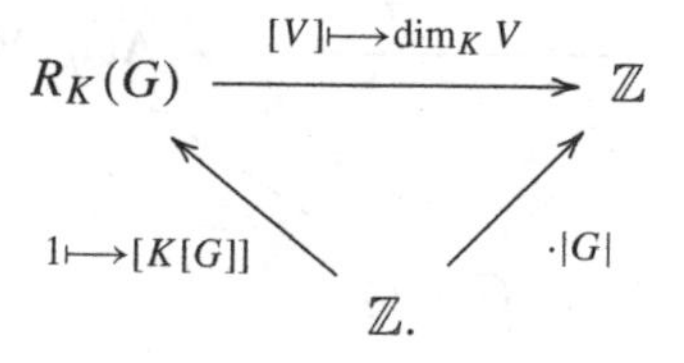

We also see that the trivial $K[G]$-module K has the G-invariant lattice R which cannot be projective as an $R[G]$-module if $G \ne \{1\}$.

Before we can establish the finer properties of the Cartan–Brauer triangle we need to develop the theory of induction.

2.3 The Ring Structure of $R_F(G)$, and Induction

In this section we let F be an arbitrary field, and we consider the group ring $F[G]$ and its Grothendieck group $R_F(G) := R(F[G])$.

Let V and W be two (finitely generated) $F[G]$-modules. The group G acts on the tensor product $V \otimes_F W$ by

$$g(v \otimes w) := gv \otimes gw \quad \text{for } v \in V \text{ and } w \in W.$$

In this way $V \otimes_F W$ becomes a (finitely generated) $F[G]$-module, and we obtain the multiplication map

$$\begin{aligned} \mathbb{Z}[\mathfrak{M}_{F[G]}] \times \mathbb{Z}[\mathfrak{M}_{F[G]}] &\longrightarrow \mathbb{Z}[\mathfrak{M}_{F[G]}] \\ (\{V\}, \{W\}) &\longmapsto \{V \otimes_F W\}. \end{aligned}$$

Since the tensor product, up to isomorphism, is associative and commutative this multiplication makes $\mathbb{Z}[\mathfrak{M}_{F[G]}]$ into a commutative ring. Its unit element is the isomorphism class $\{F\}$ of the trivial $F[G]$-module.

Remark 2.3.1 The subgroup Rel is an ideal in the ring $\mathbb{Z}[\mathfrak{M}_{F[G]}]$.

Proof Let V be a (finitely generated) $F[G]$-module and let

$$0 \longrightarrow L \xrightarrow{\alpha} M \xrightarrow{\beta} N \longrightarrow 0$$

be a short exact sequence of $F[G]$-modules. We claim that the sequence

$$0 \longrightarrow V \otimes_F L \xrightarrow{\mathrm{id}_V \otimes \alpha} V \otimes_F M \xrightarrow{\mathrm{id}_V \otimes \beta} V \otimes_F N \longrightarrow 0$$

is exact as well. This shows that the subgroup Rel is preserved under multiplication by $\{V\}$. The exactness in question is purely a problem about F-vector spaces. But as vector spaces we have $M \cong L \oplus N$ and hence $V \otimes_F M \cong (V \otimes_F L) \oplus (V \otimes_F N)$. □

It follows that $R_F(G)$ naturally is a commutative ring with unit element $[F]$ such that

$$[V] \cdot [W] = [V \otimes_F W].$$

Let $H \subseteq G$ be a subgroup. Then $F[H] \subseteq F[G]$ is a subring (with the same unit element). Any $F[G]$-module V, by restriction of scalars, can be viewed as an $F[H]$-module. If V is finitely generated as an $F[G]$-module then V is a finite-dimensional F-vector space and, in particular, is finitely generated as an $F[H]$-module. Hence we have the ring homomorphism

$$\begin{aligned} \mathrm{res}_H^G: \quad R_F(G) &\longrightarrow R_F(H) \\ [V] &\longmapsto [V]. \end{aligned}$$

On the other hand, for any $F[H]$-module W we have, by base extension, the $F[G]$-module $F[G] \otimes_{F[H]} W$. Obviously, the latter is finitely generated over $F[G]$ if the former was finitely generated over $F[H]$. The *first Frobenius reciprocity* says that

$$\begin{aligned} \mathrm{Hom}_{F[G]}\big(F[G]\otimes_{F[H]} W, V\big) &\xrightarrow{\cong} \mathrm{Hom}_{F[H]}(W, V) \\ \alpha &\longmapsto \big[w \longmapsto \alpha(1\otimes w)\big] \end{aligned}$$

is an F-linear isomorphism for any $F[H]$-module W and any $F[G]$-module V.

Remark 2.3.2 For any short exact sequence $0 \longrightarrow L \xrightarrow{\alpha} M \xrightarrow{\beta} N \longrightarrow 0$ of $F[H]$-modules the sequence of $F[G]$-modules

$$0 \to F[G]\otimes_{F[H]} L \xrightarrow{\mathrm{id}_{F[G]}\otimes\alpha} F[G]\otimes_{F[H]} M \xrightarrow{\mathrm{id}_{F[G]}\otimes\beta} F[G]\otimes_{F[H]} N \to 0$$

is exact as well.

Proof Let $g_1, \ldots, g_r \in G$ be a set of representatives for the left cosets of H in G. Then $g_1, \ldots, g_r$ also is a basis of the free right $F[H]$-module $F[G]$. It follows that, for any $F[H]$-module W, the map

$$\begin{aligned} W^r &\xrightarrow{\cong} F[G]\otimes_{F[H]} W \\ (w_1, \ldots, w_r) &\longmapsto g_1\otimes w_1 + \cdots + g_r\otimes w_r \end{aligned}$$

is an F-linear isomorphism. We see that, as a sequence of F-vector spaces, the sequence in question is just the r-fold direct sum of the original exact sequence with itself. □

Remark 2.3.3 For any $F[H]$-module W, if the $F[G]$-module $F[G]\otimes_{F[H]} W$ is simple then W is a simple $F[H]$-module.

Proof Let $W' \subseteq W$ be any $F[H]$-submodule. Then $F[G]\otimes_{F[H]} W'$ is an $F[G]$-submodule of $F[G]\otimes_{F[H]} W$ by Remark 2.3.2. Since the latter is simple we must have $F[G]\otimes_{F[H]} W' = \{0\}$ or $= F[G]\otimes_{F[H]} W$. Comparing dimensions using the argument in the proof of Remark 2.3.2 we obtain

$$\begin{aligned} [G:H]\cdot \dim_F W' = \dim_F F[G]\otimes_{F[H]} W' &= 0 \quad \text{or} \\ &= \dim_F F[G]\otimes_{F[H]} W = [G:H]\cdot \dim_F W. \end{aligned}$$

We see that $\dim_F W' = 0$ or $= \dim_F W$ and therefore that $W' = \{0\}$ or $= W$. This proves that W is a simple $F[H]$-module. □

It follows that the map

$$\begin{aligned} \mathbb{Z}[\mathfrak{M}_{F[H]}] &\longrightarrow \mathbb{Z}[\mathfrak{M}_{F[G]}] \\ \{W\} &\longmapsto \big\{F[G]\otimes_{F[H]} W\big\} \end{aligned}$$

preserves the subgroups Rel in both sides and therefore induces an additive homomorphism

$$\begin{aligned}\operatorname{ind}_H^G:\quad R_F(H) &\longrightarrow R_F(G)\\ [W] &\longmapsto \bigl[F[G]\otimes_{F[H]} W\bigr]\end{aligned}$$

(which is not multiplicative!).

Proposition 2.3.4 *We have*

$$\operatorname{ind}_H^G(y)\cdot x = \operatorname{ind}_H^G\bigl(y\cdot \operatorname{res}_H^G(x)\bigr) \quad \text{for any } x\in R_F(G) \text{ and } y\in R_F(H).$$

Proof It suffices to show that, for any $F[G]$-module V and any $F[H]$-module W, we have an isomorphism of $F[G]$-modules

$$\bigl(F[G]\otimes_{F[H]} W\bigr)\otimes_F V \cong F[G]\otimes_{F[H]} (W\otimes_F V).$$

One checks (exercise!) that such an isomorphism is given by

$$(g\otimes w)\otimes v \longmapsto g\otimes \bigl(w\otimes g^{-1}v\bigr).$$

□

Corollary 2.3.5 *The image of* $\operatorname{ind}_H^G : R_F(H) \longrightarrow R_F(G)$ *is an ideal in* $R_F(G)$.

We also mention the obvious transitivity relations

$$\operatorname{res}_{H'}^H \circ \operatorname{res}_H^G = \operatorname{res}_{H'}^G \quad \text{and} \quad \operatorname{ind}_H^G \circ \operatorname{ind}_{H'}^H = \operatorname{ind}_{H'}^G$$

for any chain of subgroups $H'\subseteq H\subseteq G$.

An alternative way to look at induction is the following. Let W be any $F[H]$-module. Then

$$\operatorname{Ind}_H^G(W) := \bigl\{\phi : G\longrightarrow W : \phi(gh) = h^{-1}\phi(g) \text{ for any } g\in G, h\in H\bigr\}$$

equipped with the *left translation action* of G given by

$${}^g\phi(g') := \phi(g^{-1}g')$$

is an $F[G]$-module called the *module induced* from W. But, in fact, the map

$$\begin{aligned}F[G]\otimes_{F[H]} W &\xrightarrow{\cong} \operatorname{Ind}_H^G(W)\\ \Bigl(\sum_{g\in G} a_g g\Bigr)\otimes w &\longmapsto \phi(g') := \sum_{h\in H} a_{g'h} h w\end{aligned}$$

is an isomorphism of $F[G]$-modules. This leads to the *second Frobenius reciprocity* isomorphism

$$\begin{aligned}\operatorname{Hom}_{F[G]}\bigl(V, \operatorname{Ind}_H^G(W)\bigr) &\xrightarrow{\cong} \operatorname{Hom}_{F[H]}(V, W)\\ \alpha &\longmapsto \bigl[v\longmapsto \alpha(v)(1)\bigr].\end{aligned}$$

We also need to recall the character theory of G in the semisimple case. For this we assume for the rest of this section that the order of G is prime to the characteristic of the field F. Any finitely generated $F[G]$-module V is a finite-dimensional F-vector space. Hence we may introduce the function

$$\begin{aligned} \chi_V\colon\quad G &\longrightarrow F \\ g &\longmapsto \operatorname{tr}(g;V) = \text{trace of } V \xrightarrow{g} V \end{aligned}$$

which is called the *character* of V. It depends only on the isomorphism class of V. Characters are class functions on G, i.e. they are constant on each conjugacy class of G. For any two finitely generated $F[G]$-modules V_1 and V_2 we have

$$\chi_{V_1 \oplus V_2} = \chi_{V_1} + \chi_{V_2} \quad \text{and} \quad \chi_{V_1 \otimes_F V_2} = \chi_{V_1} \cdot \chi_{V_2}.$$

Let $\mathrm{Cl}(G,F)$ denote the F-vector space of all class functions $G \longrightarrow F$. By pointwise multiplication of functions it is a commutative F-algebra. The above identities imply that the map

$$\begin{aligned} \mathrm{Tr}\colon\quad R_F(G) &\longrightarrow \mathrm{Cl}(G,F) \\ [V] &\longmapsto \chi_V \end{aligned}$$

is a ring homomorphism.

Definition The field F is called a splitting field for G if, for any simple $F[G]$-module V, we have $\mathrm{End}_{F[G]}(V) = F$.

If F is algebraically closed then it is a splitting field for G.

Theorem 2.3.6

i. *If the field F has characteristic zero then the characters $\{\chi_V : \{V\} \in \widehat{F[G]}\}$ are F-linearly independent.*
ii. *If F is a splitting field for G then the characters $\{\chi_V : \{V\} \in \widehat{F[G]}\}$ form a basis of the F-vector space* $\mathrm{Cl}(G,F)$.
iii. *If F has characteristic zero then two finitely generated $F[G]$-modules V_1 and V_2 are isomorphic if and only if $\chi_{V_1} = \chi_{V_2}$ holds true.*

Corollary 2.3.7

i. *If F has characteristic zero then the map* Tr *is injective.*
ii. *If F is a splitting field for G then the map* Tr *induces an isomorphism of F-algebras*

$$F \otimes_{\mathbb{Z}} R_F(G) \xrightarrow{\cong} \mathrm{Cl}(G,F).$$

iii. *If F has characteristic zero then the map*

$$\mathfrak{M}_{F[G]}/\cong \longrightarrow \mathrm{Cl}(G, F)$$
$$\{V\} \longmapsto \chi_V$$

is injective.

Proof For i. and ii. use Proposition 1.7.1. □

2.4 The Burnside Ring

A G-set X is a set equipped with a G-action

$$G \times X \longrightarrow X$$
$$(g, x) \longmapsto gx$$

such that

$$1x = x \quad \text{and} \quad g(hx) = (gh)x \quad \text{for any } g, h \in G \text{ and any } x \in X.$$

Let X and Y be two G-sets. Their disjoint union $X \cup Y$ is a G-set in an obvious way. But also their cartesian product $X \times Y$ is a G-set with respect to

$$g(x, y) := (gx, gy) \quad \text{for } (x, y) \in X \times Y.$$

We will call X and Y isomorphic if there is a bijective map $\alpha : X \xrightarrow{\simeq} Y$ such that $\alpha(gx) = g\alpha(x)$ for any $g \in G$ and $x \in X$.

Let $\mathcal{S}_G$ denote the set of all isomorphism classes $\{X\}$ of finite G-sets X. In the free abelian group $\mathbb{Z}[\mathcal{S}_G]$ we consider the subgroup Rel generated by all elements of the form

$$\{X \cup Y\} - \{X\} - \{Y\} \quad \text{for any two finite } G\text{-sets } X \text{ and } Y.$$

We define the factor group

$$B(G) := \mathbb{Z}[\mathcal{S}_G]/\mathrm{Rel},$$

and we let $[X] \in B(G)$ denote the image of the isomorphism class $\{X\}$. In fact, the map

$$\mathbb{Z}[\mathcal{S}_G] \times \mathbb{Z}[\mathcal{S}_G] \longrightarrow \mathbb{Z}[\mathcal{S}_G]$$
$$\big(\{X\}, \{Y\}\big) \longmapsto \{X \times Y\}$$

makes $\mathbb{Z}[\mathcal{S}_G]$ into a commutative ring in which the unit element is the isomorphism class of the G-set with one point. Because of $(X_1 \cup X_2) \times Y = (X_1 \times Y) \cup (X_2 \times Y)$

the subgroup Rel is an ideal in $\mathbb{Z}[\mathcal{S}_G]$. We see that $B(G)$ is a commutative ring. It is called the *Burnside ring* of G.

Two elements x, y in a G-set X are called equivalent if there is a $g \in G$ such that $y = gx$. This defines an equivalence relation on X. The equivalence classes are called *G-orbits*. They are of the form $Gx = \{gx : g \in G\}$ for some $x \in X$. A nonempty G-set which consists of a single G-orbit is called *simple* (or transitive or a principal homogeneous space). The decomposition of an arbitrary G-set X into its G-orbits is the unique decomposition of X into simple G-sets. In particular, the only G-subsets of a simple G-set Y are Y and $\emptyset$. We let $\mathcal{S}_G$ denote the set of isomorphism classes of simple G-sets.

Lemma 2.4.1 $\mathbb{Z}[\mathcal{S}_G] \xrightarrow{\cong} B(G)$.

Proof If $X = Y_1 \cup \cdots \cup Y_n$ is the decomposition of X into its G-orbits then we put

$$\pi\big(\{X\}\big) := \{Y_1\} + \cdots + \{Y_n\}.$$

This defines an endomorphism π of $\mathbb{Z}[\mathcal{S}_G]$ which is idempotent with $\mathrm{im}(\pi) = \mathbb{Z}[\mathcal{S}_G]$. It is rather clear that $\mathrm{Rel} \subseteq \ker(\pi)$. Moreover, using the identities

$$[Y_1] + [Y_2] = [Y_1 \cup Y_2], \qquad [Y_1 \cup Y_2] + [Y_3] = [Y_1 \cup Y_2 \cup Y_3], \qquad \ldots,$$
$$[Y_1 \cup \cdots \cup Y_{n-1}] + [Y_n] = [X]$$

we see that

$$[X] = [Y_1] + \cdots + [Y_n]$$

and hence that $\{X\} - \pi(\{X\}) \in \mathrm{Rel}$. As in the proof of Proposition 1.7.1 these three facts together imply

$$\mathbb{Z}[\mathcal{S}_G] = \mathbb{Z}[S_G] \oplus \mathrm{Rel}. \qquad \square$$

For any subgroup $H \subseteq G$ the coset space G/H is a simple G-set with respect to

$$\begin{aligned} G \times G/H &\longrightarrow G/H \\ \big(g, g'H\big) &\longmapsto gg'H. \end{aligned}$$

Simple G-sets of this form are called *standard G-sets*.

Remark 2.4.2 Each simple G-set X is isomorphic to some standard G-set G/H.

Proof We fix a point $x \in X$. Let $G_x := \{g \in G : gx = x\}$ be the stabilizer of x in G. Then

$$\begin{aligned} G/G_x &\xrightarrow{\simeq} X \\ gG_x &\longmapsto gx \end{aligned}$$

is an isomorphism. $\square$

It follows that the set S_G is finite.

Lemma 2.4.3 *Two standard G-sets G/H_1 and G/H_2 are isomorphic if and only if there is a $g_0 \in G$ such that $g_0^{-1} H_1 g_0 = H_2$.*

Proof If $g_0^{-1} H_1 g_0 = H_2$ then

$$\begin{aligned} G/H_1 &\xrightarrow{\simeq} G/H_2 \\ gH_1 &\longmapsto gg_0 H_2 \end{aligned}$$

is an isomorphism of G-sets. Vice versa, let

$$\alpha\colon\ G/H_1 \xrightarrow{\simeq} G/H_2$$

be an isomorphism of G-sets. We have $\alpha(1H_1) = g_0 H_2$ for some $g_0 \in G$ and then

$$g_0 H_2 = \alpha(1H_1) = \alpha(h_1 H_1) = h_1 \alpha(1H_1) = h_1 g_0 H_2$$

for any $h_1 \in H_1$. This implies $g_0^{-1} H_1 g_0 \subseteq H_2$. On the other hand

$$g_0^{-1} H_1 = \alpha^{-1}(1H_2) = \alpha^{-1}(h_2 H_2) = h_2 \alpha^{-1}(1H_2) = h_2 g_0^{-1} H_1$$

for any $h_2 \in H_2$ which implies $g_0 H_2 g_0^{-1} \subseteq H_1$. □

Exercise 2.4.4 Let G/H_1 and G/H_2 be two standard G-sets; we then have

$$[G/H_1] \cdot [G/H_2] = \sum_{g \in H_1 \backslash G/H_2} \left[G/H_1 \cap gH_2 g^{-1}\right] \quad \text{in } B(G)$$

where $H_1 \backslash G/H_2$ denotes the space of double cosets $H_1 g H_2$ in G.

Let F again be an arbitrary field. For any finite set X we have the finite-dimensional F-vector space

$$F[X] := \left\{ \sum_{x \in X} a_x x : a_x \in F \right\}$$

"with basis X". Suppose that X is a finite G-set. Then the group G acts on $F[X]$ by

$$g\left(\sum_{x \in X} a_x x\right) := \sum_{x \in X} a_x gx = \sum_{x \in X} a_{g^{-1}x} x.$$

In this way $F[X]$ becomes a finitely generated $F[G]$-module (called a *permutation module*). If $\alpha : X \xrightarrow{\simeq} Y$ is an isomorphism of finite G-sets then

$$\begin{aligned} \tilde{\alpha}\colon\quad F[X] &\xrightarrow{\cong} F[Y] \\ \sum_{x \in X} a_x x &\longmapsto \sum_{x \in X} a_x \alpha(x) = \sum_{y \in Y} a_{\alpha^{-1}(y)} y \end{aligned}$$

is an isomorphism of $F[G]$-modules. It follows that the map

$$\begin{aligned} \mathcal{S}_G &\longrightarrow \mathfrak{M}_{F[G]}/\cong \\ \{X\} &\longmapsto \{F[X]\} \end{aligned}$$

is well defined. We obviously have

$$F[X_1 \cup X_2] = F[X_1] \oplus F[X_2]$$

for any two finite G-sets X_1 and X_2. Hence the above map respects the subgroups Rel in both sides and induces a group homomorphism

$$\begin{aligned} b\colon \quad B(G) &\longrightarrow R_F(G) \\ [X] &\longmapsto [F[X]]. \end{aligned}$$

Remark There is a third interesting Grothendieck group for the ring $F[G]$ which is the factor group

$$A_F(G) := \mathbb{Z}[\mathfrak{M}_{F[G]}]/\operatorname{Rel}_\oplus$$

with respect to the subgroup $\operatorname{Rel}_\oplus$ generated by all elements of the form

$$\{M \oplus N\} - \{M\} - \{N\}$$

where M and N are arbitrary finitely generated $F[G]$-modules. We note that $\operatorname{Rel}_\oplus \subseteq \operatorname{Rel}$. The above map b is the composite of the maps

$$\begin{array}{ccccc} B(G) & \longrightarrow & A_F(G) & \xrightarrow{\text{pr}} & R_F(G) \\ [X] & \longmapsto & [F[X]] & \longmapsto & [F[X]]. \end{array}$$

Remark 2.4.5 For any two finite G-sets X_1 and X_2 we have

$$F[X_1 \times X_2] \cong F[X_1] \otimes_F F[X_2]$$

as $F[G]$-modules.

Proof The vectors (x_1, x_2), resp. $x_1 \otimes x_2$, for $x_1 \in X_1$ and $x_2 \in X_2$, form an F-basis of the left-, resp. right-, hand side. Hence there is a unique F-linear isomorphism mapping (x_1, x_2) to $x_1 \otimes x_2$. Because of

$$g(x_1, x_2) = (gx_1, gx_2) \longmapsto gx_1 \otimes gx_2 = g(x_1 \otimes x_2),$$

for any $g \in G$, this map is an $F[G]$-module isomorphism. □

It follows that the map

$$b\colon \quad B(G) \longrightarrow R_F(G)$$

is a ring homomorphism. Note that the unit element $[G/G]$ in $B(G)$ is mapped to the class $[F]$ of the trivial module F which is the unit element in $R_F(G)$.

Lemma 2.4.6

i. *For any standard G-set G/H we have*

$$b([G/H]) = \operatorname{ind}_H^G(1)$$

where 1 on the right-hand side denotes the unit element of $R_F(H)$.

ii. *For any finite G-set X we have*

$$\operatorname{tr}(g; F[X]) = |\{x \in X : gx = x\}| \in F \quad \text{for any } g \in G.$$

Proof i. Let F be the trivial $F[H]$-module. It suffices to establish an isomorphism

$$F[G/H] \cong F[G] \otimes_{F[H]} F.$$

For this purpose we consider the F-bilinear map

$$\begin{aligned} \beta\colon\quad F[G] \times F &\longrightarrow F[G/H] \\ \Big(\sum_{g\in G} a_g g, a\Big) &\longmapsto \sum_{g\in G} a a_g gH. \end{aligned}$$

Because of

$$\beta(gh, a) = ghH = gH = \beta(g, a) = \beta(g, ha)$$

for any $h \in H$ the map β is $F[H]$-balanced and therefore induces a well-defined F-linear map

$$\begin{aligned} \tilde\beta\colon\quad F[G] \otimes_{F[H]} F &\longrightarrow F[G/H] \\ \Big(\sum_{g\in G} a_g g\Big) \otimes a &\longmapsto \sum_{g\in G} a a_g gH. \end{aligned}$$

As discussed in the proof of Remark 2.3.2 we have $F[G] \otimes_{F[H]} F \cong F^{[G:H]}$ as F-vector spaces. Hence $\tilde\beta$ is a map between F-vector spaces of the same dimension. It obviously is surjective and therefore bijective. Finally the identity

$$\begin{aligned} \tilde\beta\Big(g'\Big(\Big(\sum_{g\in G} a_g g\Big) \otimes a\Big)\Big) &= \tilde\beta\Big(\Big(\sum_{g\in G} a_g g' g\Big) \otimes a\Big) = \sum_{g\in G} a a_g g' gH \\ &= g'\Big(\sum_{g\in G} a a_g gH\Big) = g'\tilde\beta\Big(\Big(\sum_{g\in G} a_g g\Big) \otimes a\Big) \end{aligned}$$

for any $g' \in G$ shows that $\tilde\beta$ is an isomorphism of $F[G]$-modules.

ii. The matrix $(a_{x,y})_{x,y}$ of the F-linear map $F[X] \xrightarrow{g\cdot} F[X]$ with respect to the basis X is given by the equations

$$gy = \sum_{x \in X} a_{x,y} x.$$

But $gy \in X$ and hence

$$a_{x,y} = \begin{cases} 1 & \text{if } x = gy \\ 0 & \text{otherwise.} \end{cases}$$

It follows that

$$\operatorname{tr}\bigl(g; F[X]\bigr) = \sum_{x \in X} a_{x,x} = \sum_{gx=x} 1 = \bigl|\{x \in X : gx = x\}\bigr| \in F.$$

□

Remark

1. The map b rarely is injective. Let $G = S_3$ be the symmetric group on three letters. It has four conjugacy classes of subgroups. Using Lemma 2.4.1, Remark 2.4.2, and Lemma 2.4.3 it therefore follows that $B(S_3) \cong \mathbb{Z}^4$. On the other hand, S_3 has only three conjugacy classes of elements. Hence Proposition 1.7.1 and Theorem 2.3.6.ii imply that $R_{\mathbb{C}}(S_3) \cong \mathbb{Z}^3$.
2. In general the map b is not surjective either. But there are many structural results about its cokernel. For example, the Artin induction theorem implies that

$$|G| \cdot R_{\mathbb{Q}}(G) \subseteq \operatorname{im}(b).$$

We therefore introduce the subring

$$P_F(G) := \operatorname{im}(b) \subseteq R_F(G).$$

Let $\mathcal{H}$ be a family of subgroups of G with the property that if $H' \subseteq H$ is a subgroup of some $H \in \mathcal{H}$ then also $H' \in \mathcal{H}$. We introduce the subgroup $B(G, \mathcal{H}) \subseteq B(G)$ generated by all $[G/H]$ for $H \in \mathcal{H}$ as well as its image $P_F(G, \mathcal{H}) \subseteq P_F(G)$ under the map b.

Lemma 2.4.7 *$B(G, \mathcal{H})$ is an ideal in $B(G)$, and hence $P_F(G, \mathcal{H})$ is an ideal in $P_F(G)$.*

Proof We have to show that, for any $H_1 \in \mathcal{H}$ and any subgroup $H_2 \subseteq G$, the element $[G/H_1] \cdot [G/H_2]$ lies in $B(G, \mathcal{H})$. This is immediately clear from Exercise 2.4.4. But a less detailed argument suffices. Obviously $[G/H_1] \cdot [G/H_2]$ is the sum of the classes of the G-orbits in $G/H_1 \times G/H_2$. Let $G(g_1H_1, g_2H_2) = G(H_1, g_1^{-1}g_2H_2)$ be such a G-orbit. The stabilizer $H' \subseteq G$ of the element $(H_1, g_1^{-1}g_2H_2)$ is contained in H_1 and therefore belongs to $\mathcal{H}$. It follows that

$$\bigl[G(g_1H_1, g_2H_2)\bigr] = \bigl[G/H'\bigr] \in B(G, \mathcal{H}).$$

□

Definition

i. Let ℓ be a prime number. A finite group H is called ℓ-hyper-elementary if it contains a cyclic normal subgroup C such that $\ell \nmid |C|$ and H/C is an ℓ-group.
ii. A finite group is called hyper-elementary if it is ℓ-hyper-elementary for some prime number ℓ.

Exercise Let H be an ℓ-hyper-elementary group. Then:

i. Any subgroup of H is ℓ-hyper-elementary;
ii. let $C \subseteq H$ be a cyclic normal subgroup as in the definition, and let $L \subseteq H$ be any ℓ-Sylow subgroup; then the map $C \times L \xrightarrow{\cong} H$ sending (c, g) to cg is a bijection of sets.

Let $\mathcal{H}_{he}$ denote the family of hyper-elementary subgroups of G. By the exercise Lemma 2.4.7 is applicable to $\mathcal{H}_{he}$.

Theorem 2.4.8 (Solomon) *Suppose that F has characteristic zero; then*

$$P_F(G, \mathcal{H}_{he}) = P_F(G).$$

Proof Because of Lemma 2.4.7 it suffices to show that the unit element $1 \in P_F(G)$ already lies in $P_F(G, \mathcal{H}_{he})$. According to Lemma 2.4.6.ii the characters

$$\chi_{F[X]}(g) = \big|\{x \in X : gx = x\}\big| \in \mathbb{Z},$$

for any $g \in G$ and any finite G-set X, have integral values. Hence we have the well-defined ring homomorphisms

$$\begin{aligned} t_g\colon \ P_F(G) &\longrightarrow \mathbb{Z} \\ z &\longmapsto \operatorname{Tr}(z)(g) \end{aligned}$$

for $g \in G$. On the one hand, by Corollary 2.3.7.i, they satisfy

$$\bigcap_{g \in G} \ker(t_g) = \ker\big(\operatorname{Tr} \mid P_F(G)\big) = \{0\}. \tag{2.4.1}$$

On the other hand, we claim that

$$t_g\big(P_F(G, \mathcal{H}_{he})\big) = \mathbb{Z} \tag{2.4.2}$$

holds true for any $g \in G$. We fix a $g_0 \in G$ in the following. Since the image $t_{g_0}(P_F(G, \mathcal{H}_{he}))$ is an additive subgroup of $\mathbb{Z}$ and hence is of the form $n\mathbb{Z}$ for some $n \geq 0$ it suffices to find, for any prime number ℓ, an ℓ-hyper-elementary subgroup $H \subseteq G$ such that the integer

$$t_{g_0}([F[G/H]]) = \chi_{F[G/H]}(g_0) = |\{x \in G/H : g_0 x = x\}|$$
$$= |\{gH \in G/H : g^{-1} g_0 g \in H\}|$$

is not contained in $\ell\mathbb{Z}$. We also fix ℓ.

The wanted ℓ-hyper-elementary subgroup H will be found in a chain of subgroups

$$C \subseteq \langle g_0 \rangle \subseteq H \subseteq N$$

with C being normal in N which is constructed as follows. Let $n \geq 1$ be the order of g_0, and write $n = \ell^s m$ with $l \nmid m$. The cyclic subgroup $\langle g_0 \rangle \subseteq G$ generated by g_0 then is the direct product

$$\langle g_0 \rangle = \langle g_0^{\ell^s} \rangle \times \langle g_0^m \rangle$$

where $\langle g_0^m \rangle$ is an ℓ-group and $C := \langle g_0^{\ell^s} \rangle$ is a cyclic group of order prime to ℓ. We define $N := \{g \in G : gCg^{-1} = C\}$ to be the normalizer of C in G. It contains $\langle g_0 \rangle$, of course. Finally, we choose $H \subseteq N$ in such a way that H/C is an ℓ-Sylow subgroup of N/C which contains the ℓ-subgroup $\langle g_0 \rangle / C$. By construction H is ℓ-hyper-elementary.

In the next step we study the cardinality of the set

$$\{gH \in G/H : g_0 g H = gH\} = \{gH \in G/H : g^{-1} g_0 g \in H\}.$$

Suppose that $g^{-1} g_0 g \in H$. Then $g^{-1} C g \subseteq g^{-1} \langle g_0 \rangle g = \langle g^{-1} g_0 g \rangle \subseteq H$. But, the two sides having coprime orders, the projection map $g^{-1} C g \longrightarrow H/C$ has to be the trivial map. It follows that $g^{-1} C g = C$ which means that $g \in N$. This shows that

$$\{gH \in G/H : g_0 g H = gH\} = \{gH \in N/H : g_0 g H = gH\}.$$

The cardinality of the right-hand side is the number of $\langle g_0 \rangle$-orbits in N/H which consist of one point only. We note that the subgroup C, being normal in N and contained in H, acts trivially on N/H. Hence the $\langle g_0 \rangle$-orbits coincide with the orbits of the ℓ-group $\langle g_0 \rangle / C$. But, quite generally, the cardinality of an orbit, being the index of the stabilizer of any point in the orbit, divides the order of the acting group. It follows that the cardinality of any $\langle g_0 \rangle$-orbit in N/H is a power of ℓ. We conclude that

$$|\{gH \in N/H : g_0 g H = gH\}| \equiv |N/H| = [N : H] \bmod \ell.$$

But by the choice of H we have $\ell \nmid [N : H]$. This establishes our claim.

By (2.4.2) we now may choose an element $z_g \in P_F(G, \mathcal{H}_{he})$, for any $g \in G$, such that $t_g(z_g) = 1$. We then have

$$t_g\Big(\prod_{g' \in G} (z_{g'} - 1)\Big) = 0 \quad \text{for any } g \in G,$$

and (2.4.1) implies that

$$\prod_{g'\in G}(z_{g'}-1)=0.$$

Multiplying out the left-hand side and using that $P_F(G,\mathcal{H}_{he})$ is additively and multiplicatively closed easily shows that $1\in P_F(G,\mathcal{H}_{he})$. □

2.5 Clifford Theory

As before F is an arbitrary field. We fix a normal subgroup N in our finite group G. Let W be an $F[N]$-module. It is given by a homomorphism of F-algebras

$$\pi:\quad F[N]\longrightarrow \mathrm{End}_F(W).$$

For any $g\in G$ we now define a new $F[N]$-module $g^*(W)$ by the composite homomorphism

$$\begin{aligned} F[N] &\longrightarrow F[N] \xrightarrow{\pi} \mathrm{End}_F(W)\\ h &\longmapsto ghg^{-1} \end{aligned}$$

or equivalently by

$$\begin{aligned} F[N]\times g^*(W) &\longrightarrow g^*(W)\\ (h,w) &\longmapsto ghg^{-1}w. \end{aligned}$$

Remark 2.5.1

i. $\dim_F g^*(W)=\dim_F W$.
ii. The map $U\mapsto g^*(U)$ is a bijection between the set of $F[N]$-submodules of W and the set of $F[N]$-submodules of $g^*(W)$.
iii. W is simple if and only if $g^*(W)$ is simple.
iv. $g_1^*(g_2^*(W))=(g_2g_1)^*(W)$ for any $g_1,g_2\in G$.
v. Any $F[N]$-module homomorphism $\alpha: W_1\longrightarrow W_2$ also is a homomorphism of $F[N]$-modules $\alpha: g^*(W_1)\longrightarrow g^*(W_2)$.

Proof Trivial or straightforward. □

Suppose that $g\in N$. One checks that then

$$\begin{aligned} W &\xrightarrow{\cong} g^*(W)\\ w &\longmapsto gw \end{aligned}$$

is an isomorphism of $F[N]$-modules. Together with Remark 2.5.1.iv/v this implies that

$$\begin{aligned} G/N \times (\mathfrak{M}_{F[N]}/\cong) &\longrightarrow \mathfrak{M}_{F[N]}/\cong \\ (gN, \{W\}) &\longmapsto \{g^*(W)\} \end{aligned}$$

is a well-defined action of the group G/N on the set $\mathfrak{M}_{F[N]}/\cong$. By Remark 2.5.1.iii this action respects the subset $\widehat{F[N]}$. For any $\{W\}$ in $\widehat{F[N]}$ we put

$$I_G(W) := \{g \in G : \{g^*(W)\} = \{W\}\}.$$

As a consequence of Remark 2.5.1.iv this is a subgroup of G, which contains N of course.

Remark 2.5.2 Let V be an $F[G]$-module, and let $g \in G$ be any element; then the map

$$\begin{aligned} \begin{array}{c}\text{set of all } F[N]\text{-}\\ \text{submodules of } V\end{array} &\xrightarrow{\cong} \begin{array}{c}\text{set of all } F[N]\text{-}\\ \text{submodules of } V\end{array} \\ W &\longmapsto gW \end{aligned}$$

is an inclusion preserving bijection; moreover, for any $F[N]$-submodule $W \subseteq V$ we have:

i. The map

$$\begin{aligned} g^*(W) &\xrightarrow{\cong} g^{-1}W \\ w &\longmapsto g^{-1}w \end{aligned}$$

is an isomorphism of $F[N]$-modules;
ii. gW is a simple $F[N]$-module if and only if W is a simple $F[N]$-module;
iii. if $W_1 \cong W_2$ are isomorphic $F[N]$-submodules of V then also $gW_1 \cong gW_2$ are isomorphic as $F[N]$-modules.

Proof For $h \in N$ we have

$$h(gW) = g(g^{-1}hg)W = gW$$

since $g^{-1}hg \in N$. Hence gW indeed is an $F[N]$-submodule, and the map in the assertion is well defined. It obviously is inclusion preserving. Its bijectivity is immediate from $g^{-1}(gW) = W = g(g^{-1}W)$. The assertion ii. is a direct consequence. The map in i. clearly is an F-linear isomorphism. Because of

$$g^{-1}(ghg^{-1}w) = h(g^{-1}w) \quad \text{for any } h \in N \text{ and } w \in W$$

it is an $F[N]$-module isomorphism. The last assertion iii. follows from i. and Remark 2.5.1.v. □

Theorem 2.5.3 (Clifford) *Let V be a simple $F[G]$-module; we then have:*

i. *V is semisimple as an $F[N]$-module;*
ii. *let $W \subseteq V$ be a simple $F[N]$-submodule, and let $\tilde{W} \subseteq V$ be the $\{W\}$-isotypic component; then*
   a. *$\tilde{W}$ is a simple $F[I_G(W)]$-module, and*
   b. *$V \cong \mathrm{Ind}^G_{I_G(W)}(\tilde{W})$ as $F[G]$-modules.*

Proof Since V is of finite length as an $F[N]$-module we find a simple $F[N]$-submodule $W \subseteq V$. Then gW, for any $g \in G$, is another simple $F[N]$-submodule by Remark 2.5.2.ii. Therefore $V_0 := \sum_{g\in G} gW$ is, on the one hand, a semisimple $F[N]$-module by Proposition 1.1.4. On the other hand it is, by definition, a nonzero $F[G]$-submodule of V. Since V is simple we must have $V_0 = V$, which proves the assertion i. As an $F[N]$-submodule the $\{W\}$-isotypic component $\tilde{W}$ of V is of the form $\tilde{W} = W_1 \oplus \cdots \oplus W_m$ with simple $F[N]$-submodules $W_i \cong W$. Let first g be an element in $I_G(W)$. Using Remark 2.5.2.i/iii we obtain $gW_i \cong gW \cong W$ for any $1 \le i \le m$. It follows that $gW_i \subseteq \tilde{W}$ for any $1 \le i \le m$ and hence $g\tilde{W} \subseteq \tilde{W}$. We see that $\tilde{W}$ is an $F[I_G(W)]$-submodule of V. For a general $g \in G$ we conclude from Remark 2.5.2 that $g\tilde{W}$ is the $\{gW\}$-isotypic component of V. We certainly have

$$V = \sum_{g\in G/I_G(W)} g\tilde{W}.$$

Two such submodules $g_1\tilde{W}$ and $g_2\tilde{W}$, being isotypic components, either are equal or have zero intersection. If $g_1\tilde{W} = g_2\tilde{W}$ then $g_2^{-1}g_1\tilde{W} = \tilde{W}$, hence $g_2^{-1}g_1W \cong W$, and therefore $g_2^{-1}g_1 \in I_G(W)$. We see that in fact

$$V = \bigoplus_{g\in G/I_G(W)} g\tilde{W}. \tag{2.5.1}$$

The inclusion $\tilde{W} \subseteq V$ induces, by the first Frobenius reciprocity, the $F[G]$-module homomorphism

$$\mathrm{Ind}^G_{I_G(W)}(\tilde{W}) \cong F[G] \otimes_{F[I_G(W)]} \tilde{W} \longrightarrow V$$

$$\Big(\sum_{g\in G} a_g g\Big) \otimes \tilde{w} \longmapsto \sum_{g\in G} a_g g\tilde{w}.$$

Since V is simple it must be surjective. But both sides have the same dimension as F-vector spaces $[G : I_G(W)] \cdot \dim_F \tilde{W}$, the left-hand side by the argument in the proof of Remark 2.3.2 and the right-hand side by (2.5.1). Hence this map is an isomorphism which proves ii.b. Finally, since $F[G] \otimes_{F[I_G(W)]} \tilde{W} \cong V$ is a simple $F[G]$-module it follows from Remark 2.3.3 that $\tilde{W}$ is a simple $F[I_G(W)]$-module. □

In the next section we will need the following particular consequence of this result. But first we point out that for an $F[N]$-module W of dimension $\dim_F W = 1$

the describing algebra homomorphism π is of the form

$$\pi:\quad F[N]\longrightarrow F.$$

The corresponding homomorphism for $g^*(W)$ then is $x\mapsto\pi(gxg^{-1})$. Since an endomorphism of a one-dimensional F-vector space is given by multiplication by a scalar we have $g\in I_G(W)$ if and only if there is a scalar $a\in F^\times$ such that

$$a\pi(h)w=\pi\bigl(ghg^{-1}\bigr)aw\quad\text{for any } h\in N \text{ and } w\in W.$$

It follows that

$$I_G(W)=\bigl\{g\in G:\pi\bigl(ghg^{-1}\bigr)=\pi(h)\text{ for any } h\in N\bigr\}\tag{2.5.2}$$

if $\dim_F W=1$.

Remark 2.5.4 Suppose that N is abelian and that F is a splitting field for N; then any simple $F[N]$-module W has dimension $\dim_F W=1$.

Proof Let $\pi:F[N]\longrightarrow \mathrm{End}_F(W)$ be the algebra homomorphism describing W. Since N is abelian we have $\mathrm{im}(\pi)\subseteq \mathrm{End}_{F[N]}(W)$. By our assumption on the field F the latter is equal to F. This means that any element in $F[N]$ acts on W by multiplication by a scalar in F. Since W is simple this forces it to be one-dimensional. □

Proposition 2.5.5 *Let H be an ℓ-hyper-elementary group with cyclic normal subgroup C such that $\ell\nmid|C|$ and H/C is an ℓ-group, and let V be a simple $F[H]$-module; we suppose that*

a. *F is a splitting field for C,*
b. *V does not contain the trivial $F[C]$-module, and*
c. *the subgroup $C_0:=\{c\in C:cg=gc$ for any $g\in H\}$ acts trivially on V;*

then there exists a proper subgroup $H'\subsetneqq H$ and an $F[H']$-module V' such that $V\cong \mathrm{Ind}_{H'}^H(V')$ as $F[H]$-modules.

Proof We pick any simple $F[C]$-submodule $W\subseteq V$. By applying Clifford's Theorem 2.5.3 to the normal subgroup C and the module W it suffices to show that

$$I_H(W)\neq H.$$

According to Remark 2.5.4 the module W is one-dimensional and given by an algebra homomorphism $\pi:F[C]\longrightarrow F$, and by (2.5.2) we have

$$I_H(W)=\bigl\{g\in H:\pi\bigl(gcg^{-1}\bigr)=\pi(c)\text{ for any } c\in C\bigr\}.$$

The assumption c. means that

$$C_0\subseteq C_1:=\ker(\pi|C).$$

We immediately note that any subgroup of the cyclic normal subgroup C also is normal in H. By assumption b. we find an element $c_2 \in C \setminus C_1$ so that $\pi(c_2) \neq 1$. Let $L \subseteq H$ be an ℓ-Sylow subgroup. We claim that we find an element $g_0 \in L$ such that $\pi(g_0 c_2 g_0^{-1}) \neq \pi(c_2)$. Then $g_0 \notin I_H(W)$ which establishes what we wanted. We point out that, since $H = C \times L$ as sets, we have

$$C_0 = \{c \in C : cg = gc \text{ for any } g \in L\}.$$

Arguing by contradiction we assume that $\pi(gc_2g^{-1}) = \pi(c_2)$ for any $g \in L$. Then

$$gc_2C_1g^{-1} = gc_2g^{-1}C_1 = c_2C_1 \quad \text{for any } g \in L.$$

This means that we may consider c_2C_1 as an L-set with respect to the conjugation action. Since L is an ℓ-group the cardinality of any L-orbit in c_2C_1 is a power of ℓ. On the other hand we have $\ell \nmid |C_1| = |c_2C_1|$. There must therefore exist an element $c_0 \in c_2C_1$ such that

$$gc_0g^{-1} = c_0 \quad \text{for any } g \in L$$

(i.e. an L-orbit of cardinality one). We conclude that $c_0 \in C_0 \subseteq C_1$ and hence $c_2C_1 = c_0C_1 = C_1$. This is in contradiction to $c_2 \notin C_1$. □

2.6 Brauer's Induction Theorem

In this section F is a field of characteristic zero, and G continues to be any finite group.

Definition

i. Let ℓ be a prime number. A finite group H is called ℓ-elementary if it is a direct product $H = C \times L$ of a cyclic group C and an ℓ-group L.
ii. A finite group is called elementary if it is ℓ-elementary for some prime number ℓ.

Exercise

i. If H is ℓ-elementary then $H = C \times L$ is the direct product of a cyclic group C of order prime to ℓ and an ℓ-group L.
ii. Any ℓ-elementary group is ℓ-hyper-elementary.
iii. Any subgroup of an ℓ-elementary group is ℓ-elementary.

Let $\mathcal{H}_e$ denote the family of elementary subgroups of G.

Theorem 2.6.1 (Brauer) *Suppose that F is a splitting field for every subgroup of G; then*

$$\sum_{H \in \mathcal{H}_e} \operatorname{ind}_H^G\big(R_F(H)\big) = R_F(G).$$

Proof By Corollary 2.3.5 each $\mathrm{ind}_H^G(R_F(H))$ is an ideal in the ring $R_F(G)$. Hence the left-hand side of the asserted identity is an ideal in the right-hand side. To obtain equality we therefore need only to show that the unit element $1_G \in R_F(G)$ lies in the left-hand side. According to Solomon's Theorem 2.4.8 together with Lemma 2.4.6.i we have

$$1_G \in \sum_{H\in\mathcal{H}_{he}} \mathbb{Z}\big[F[G/H]\big] = \sum_{H\in\mathcal{H}_{he}} \mathbb{Z}b\big([G/H]\big) = \sum_{H\in\mathcal{H}_{he}} \mathbb{Z}\,\mathrm{ind}_H^G(1_H)$$
$$\subseteq \sum_{H\in\mathcal{H}_{he}} \mathrm{ind}_H^G\big(R_F(H)\big).$$

By the transitivity of induction this reduces us to the case that G is ℓ-hyper-elementary for some prime number ℓ. We now proceed by induction with respect to the order of G and assume that our assertion holds for all proper subgroups $H' \subsetneqq G$. We also may assume, of course, that G is not elementary. Using the transitivity of induction again it then suffices to show that

$$1_G \in \sum_{H'\subsetneqq G} \mathrm{ind}_{H'}^G\big(R_F(H')\big).$$

Let $C \subseteq G$ be the cyclic normal subgroup of order prime to ℓ such that G/C is an ℓ-group. We fix an ℓ-Sylow subgroup $L \subseteq G$. Then $G = C \times L$ as sets. In C we have the (cyclic) subgroup

$$C_0 := \{c \in C : cg = gc \text{ for any } g \in G\}.$$

Then

$$H_0 := C_0 \times L$$

is an ℓ-elementary subgroup of G. Since G is not elementary we must have $H_0 \subsetneqq G$. We consider the induction $\mathrm{Ind}_{H_0}^G(F)$ of the trivial $F[H_0]$-module F. By semisimplicity it decomposes into a direct sum

$$\mathrm{Ind}_{H_0}^G(F) = V_0 \oplus V_1 \oplus \cdots \oplus V_r$$

of simple $F[G]$-modules V_i. We recall that $\mathrm{Ind}_{H_0}^G(F)$ is the F-vector space of all functions $\phi : G/H_0 \longrightarrow F$ with the G-action given by

$${}^g\phi\big(g'H_0\big) = \phi\big(g^{-1}g'H_0\big).$$

This G-action fixes a function ϕ if and only if $\phi(gg'H_0) = \phi(g'H_0)$ for any $g, g' \in G$, i.e. if and only if ϕ is constant. It follows that the one-dimensional subspace of constant functions is the only simple $F[G]$-submodule in $\mathrm{Ind}_{H_0}^G(F)$ which is isomorphic to the trivial module. We may assume that V_0 is this trivial submodule. In $R_F(G)$ we then have the equation

$$1_G = \mathrm{ind}_{H_0}^G(1_{H_0}) - [V_1] - \cdots - [V_r].$$

This reduces us further to showing that, for any $1 \leq i \leq r$, we have

$$[V_i] \in \mathrm{ind}_{H_i}^G\big(R_F(H_i)\big)$$

for some proper subgroup $H_i \subsetneqq G$. This will be achieved by applying the criterion in Proposition 2.5.5. By our assumption on F it remains to verify the conditions b. and c. in that proposition for each $V_1, \ldots, V_r$. Since C_0 is central in G and is contained in H_0 it acts trivially on $\mathrm{Ind}_{H_0}^G(F)$ and *a fortiori* on any V_i. This is condition c. For b. we note that $CH_0 = G$ and $C \cap H_0 = C_0$. Hence the inclusion $C \subseteq G$ induces a bijection $C/C_0 \xrightarrow{\cong} G/H_0$. It follows that the map

$$\begin{aligned} \mathrm{Ind}_{H_0}^G(F) &\xrightarrow{\cong} \mathrm{Ind}_{C_0}^C(F) \\ \phi &\longmapsto \phi|(C/C_0) \end{aligned}$$

is an isomorphism of $F[C]$-modules. It maps constant functions to constant functions and hence the unique trivial $F[G]$-submodule V_0 to the unique trivial $F[C]$-submodule. Therefore V_i, for $1 \leq i \leq r$, cannot contain any trivial $F[C]$-submodule. □

Lemma 2.6.2 *Let H be an elementary group, and let $N_0 \subseteq H$ be a normal subgroup such that H/N_0 is not abelian; then there exists a normal subgroup $N_0 \subseteq N \subseteq H$ such that N/N_0 is abelian but is not contained in the center $Z(H/N_0)$ of H/N_0.*

Proof With H also H/N_0 is elementary (if $H = C \times L$ with C and L having coprime orders then $H/N_0 \cong C/C \cap N_0 \times L/L \cap N_0$). We therefore may assume without loss of generality that $N_0 = \{1\}$. *Step 1:* We assume that H is an ℓ-group for some prime number ℓ. By assumption we have $Z(H) \neq H$ so that $H/Z(H)$ is an ℓ-group $\neq\{1\}$. We pick a cyclic normal subgroup $\{1\} \neq \overline{N} = \langle\overline{g}\rangle \subseteq H/Z(H)$. Let $Z(H) \subseteq N \subseteq H$ be the normal subgroup such that $N/Z(H) = \overline{N}$ and let $g \in N$ be a preimage of $\overline{g}$. Clearly $N = \langle Z(H), g\rangle$ is abelian. But $Z(H) \subsetneqq N$ since $\overline{g} \neq 1$. *Step 2:* In general let $H = C \times L$ where C is cyclic and L is an ℓ-group. With H also L is not abelian. Applying Step 1 to L we find a normal abelian subgroup $N_L \subseteq L$ such that $N_L \nsubseteq Z(L)$. Then $N := C \times N_L$ is a normal abelian subgroup of H such that $N \nsubseteq Z(H) = C \times Z(L)$. □

Lemma 2.6.3 *Let H be an elementary group, and let W be a simple $F[H]$-module; we suppose that F is a splitting field for all subgroups of H; then there exists a subgroup $H' \subseteq H$ and a one-dimensional $F[H']$-module W' such that*

$$W \cong \mathrm{Ind}_{H'}^H(W')$$

as $F[H]$-modules.

Proof We choose $H' \subseteq H$ to be a minimal subgroup (possibly equal to H) such that there exists an $F[H']$-module W' with $W \cong \mathrm{Ind}_{H'}^{H}(W')$, and we observe that W' necessarily is a simple $F[H']$-module by Remark 2.3.3. Let

$$\pi' : \quad H' \longrightarrow \mathrm{End}_F\big(W'\big)$$

be the corresponding algebra homomorphism, and put $N_0 := \ker(\pi')$. We claim that H'/N_0 is abelian. Suppose otherwise. Then, by Lemma 2.6.2, there exists a normal subgroup $N_0 \subseteq N \subseteq H'$ such that N/N_0 is abelian but is not contained in $Z(H'/N_0)$. Let $\bar{W} \subseteq W'$ be a simple $F[N]$-submodule. By Clifford's Theorem 2.5.3 we have

$$W' \cong \mathrm{Ind}_{I_{H'}(\bar{W})}^{H'}(\tilde{\bar{W}})$$

where $\tilde{\bar{W}}$ denotes the $\{\bar{W}\}$-isotypic component of W'. Transitivity of induction implies

$$W \cong \mathrm{Ind}_{H'}^{H}\big(\mathrm{Ind}_{I_{H'}(\bar{W})}^{H'}(\tilde{\bar{W}})\big) \cong \mathrm{Ind}_{I_{H'}(\bar{W})}^{H}(\tilde{\bar{W}}).$$

By the minimality of H' we therefore must have $I_{H'}(\bar{W}) = H'$ which means that $W' = \tilde{\bar{W}}$ is $\{\bar{W}\}$-isotypic.

On the other hand, W' is an $F[H'/N_0]$-module. Hence $\bar{W}$ is a simple $F[N/N_0]$-module for the abelian group N/N_0. Remark 2.5.4 then implies (note that $\mathrm{End}_{F[N/N_0]}(\bar{W}) = \mathrm{End}_{F[N]}(\bar{W}) = F$) that $\bar{W}$ is one-dimensional given by an algebra homomorphism

$$\chi : \quad F[N/N_0] \longrightarrow F.$$

It follows that any $h \in N$ acts on the $\{\bar{W}\}$-isotypic module W' by multiplication by the scalar $\chi(hN_0)$. In other words the injective homomorphism

$$\begin{aligned} H'/N_0 &\longrightarrow \mathrm{End}_F\big(W'\big) \\ hN_0 &\longmapsto \pi'(h) \end{aligned}$$

satisfies

$$\pi'(h) = \chi(hN_0) \cdot \mathrm{id}_{W'} \quad \text{for any } h \in N.$$

But $\chi(hN_0) \cdot \mathrm{id}_{W'}$ lies in the center of $\mathrm{End}_F(W')$. The injectivity of the homomorphism therefore implies that N/N_0 lies in the center of H'/N_0. This is a contradiction.

We thus have established that H'/N_0 is abelian. Applying Remark 2.5.4 to W' viewed as a simple $F[H'/N_0]$-module we conclude that W' is one-dimensional. □

Theorem 2.6.4 (Brauer) *Suppose that F is a splitting field for any subgroup of G, and let $x \in R_F(G)$ be any element; then there exist integers $m_1, \ldots, m_r$, elementary* *subgroups $H_1, \ldots, H_r$, and one-dimensional $F[H_i]$-modules W_i such that*

$$x = \sum_{i=1}^{r} m_i \operatorname{ind}_{H_i}^{G}([W_i]).$$

Proof Combine Theorem 2.6.1, Proposition 1.7.1, Lemma 2.6.3, and the transitivity of induction. □

2.7 Splitting Fields

Again F is a field of characteristic zero.

Lemma 2.7.1 *Let E/F be any extension field, and let V and W be two finitely generated $F[G]$-modules; we then have*

$$\operatorname{Hom}_{E[G]}(E \otimes_F V, E \otimes_F W) = E \otimes_F \operatorname{Hom}_{F[G]}(V, W).$$

Proof First of all we observe, by comparing dimensions, that

$$\operatorname{Hom}_E(E \otimes_F V, E \otimes_F W) = E \otimes_F \operatorname{Hom}_F(V, W)$$

holds true. We now consider $U := \operatorname{Hom}_F(V, W)$ as an $F[G]$-module via

$$G \times U \longrightarrow U$$

$$(g, f) \longmapsto {}^g f := gf(g^{-1}).$$

Then $\operatorname{Hom}_{F[G]}(V, W) = U^G := \{f \in U : {}^g f = f \text{ for any } g \in G\}$ is the $\{F\}$-isotypic component of U for the trivial $F[G]$-module F. Correspondingly we obtain

$$\operatorname{Hom}_{E[G]}(E \otimes_F V, E \otimes_F W) = \operatorname{Hom}_E(E \otimes_F V, E \otimes_F W)^G = (E \otimes_F U)^G.$$

This reduces us to proving that

$$(E \otimes_F U)^G = E \otimes_F U^G$$

for any $F[G]$-module U. The element

$$\varepsilon_G := \frac{1}{|G|} \sum_{g \in G} g \in F[G] \subseteq E[G]$$

is an idempotent with the property that

$$U^G = \varepsilon_G \cdot U,$$

and hence

$$(E \otimes_F U)^G = \varepsilon_G \cdot (E \otimes_F U) = E \otimes_F \varepsilon_G \cdot U = E \otimes_F U^G.$$

□

Theorem 2.7.2 (Brauer) *Let e be the exponent of G, and suppose that F contains a primitive eth root of unity; then F is a splitting field for any subgroup of G.*

Proof We fix an algebraic closure $\bar{F}$ of F. *Step 1:* We show that, for any finitely generated $\bar{F}[G]$-module $\bar{V}$, there is an $F[G]$-module V such that

$$\bar{V} \cong \bar{F} \otimes_F V \quad \text{as } \bar{F}[G]\text{-modules.}$$

According to Brauer's Theorem 2.6.4 we find integers $m_1, \ldots, m_r$, subgroups $H_1, \ldots, H_r$ of G, and one-dimensional $\bar{F}[H_i]$-modules $\bar{W}_i$ such that

$$\begin{aligned}
[\bar{V}] &= \sum_{i=1}^{r} m_i \big[\bar{F}[G] \otimes_{\bar{F}[H_i]} \bar{W}_i\big] \\
&= \sum_{m_i > 0} \big[\bar{F}[G] \otimes_{\bar{F}[H_i]} \bar{W}_i^{\,m_i}\big] - \sum_{m_i < 0} \big[\bar{F}[G] \otimes_{\bar{F}[H_i]} \bar{W}_i^{\,-m_i}\big] \\
&= \Big[\bigoplus_{m_i > 0} \big(\bar{F}[G] \otimes_{\bar{F}[H_i]} \bar{W}_i^{\,m_i}\big)\Big] - \Big[\bigoplus_{m_i < 0} \big(\bar{F}[G] \otimes_{\bar{F}[H_i]} \bar{W}_i^{\,-m_i}\big)\Big].
\end{aligned}$$

Let $\pi_i : \bar{F}[H_i] \longrightarrow \bar{F}$ denote the $\bar{F}$-algebra homomorphism describing $\bar{W}_i$. We have $\pi_i(h)^e = \pi_i(h^e) = \pi_i(1) = 1$ for any $h \in H_i$. Our assumption on F therefore implies that $\pi_i(F[H_i]) \subseteq F$. Hence the restriction $\pi_i | F[H_i]$ describes a one-dimensional $F[H_i]$-module W_i such that

$$\bar{W}_i \cong \bar{F} \otimes_F W_i \quad \text{as } \bar{F}[H_i]\text{-modules.}$$

We define the $F[G]$-modules

$$V_+ := \bigoplus_{m_i > 0} \big(F[G] \otimes_{F[H_i]} W_i^{m_i}\big) \quad \text{and} \quad V_- := \bigoplus_{m_i < 0} \big(F[G] \otimes_{F[H_i]} W_i^{-m_i}\big).$$

Then

$$\begin{aligned}
\bar{F} \otimes_F V_+ &= \bigoplus_{m_i > 0} \big(\bar{F} \otimes_F F[G] \otimes_{F[H_i]} W_i^{m_i}\big) = \bigoplus_{m_i > 0} \big(\bar{F}[G] \otimes_{F[H_i]} W_i^{m_i}\big) \\
&= \bigoplus_{m_i > 0} \big(\bar{F}[G] \otimes_{\bar{F}[H_i]} \bar{F}[H_i] \otimes_{F[H_i]} W_i^{m_i}\big)
\end{aligned}$$

$$= \bigoplus_{m_i>0} \big(\bar{F}[G] \otimes_{\bar{F}[H_i]} (\bar{F} \otimes_F W_i^{m_i})\big)$$

$$\cong \bigoplus_{m_i>0} \big(\bar{F}[G] \otimes_{\bar{F}[H_i]} \bar{W_i}^{m_i}\big)$$

and similarly

$$\bar{F} \otimes_F V_- \cong \bigoplus_{m_i<0} \big(\bar{F}[G] \otimes_{\bar{F}[H_i]} \bar{W_i}^{-m_i}\big).$$

It follows that

$$[\bar{V}] = [\bar{F} \otimes_F V_+] - [\bar{F} \otimes_F V_-]$$

or, equivalently, that

$$\big[\bar{V} \oplus (\bar{F} \otimes_F V_-)\big] = [\bar{F} \otimes_F V_+].$$

Using Corollary 2.3.7.i/iii we deduce that

$$\bar{V} \oplus (\bar{F} \otimes_F V_-) \cong \bar{F} \otimes_F V_+ \quad \text{as } \bar{F}[G]\text{-modules.}$$

If V_- is nonzero then $V_- = U \oplus V_-'$ is a direct sum of $F[G]$-modules where U is simple. On the other hand let $V_+ = U_1 \oplus \cdots \oplus U_m$ be a decomposition into simple $F[G]$-modules. Then $\bar{F} \otimes_F U$ is a direct summand of $\bar{F} \otimes_F V_-$ and hence is isomorphic to a direct summand of $\bar{F} \otimes_F V_+$. We therefore must have $\mathrm{Hom}_{\bar{F}[G]}(\bar{F} \otimes_F U, \bar{F} \otimes_F U_j) \neq \{0\}$ for some $1 \leq j \leq m$. Lemma 2.7.1 implies that $\mathrm{Hom}_{F[G]}(U, U_j) \neq \{0\}$. Hence $U \cong U_j$ as $F[G]$-modules. We conclude that $V_+ \cong U \oplus V_+'$ with $V_+' := \bigoplus_{i \neq j} U_i$, and we obtain

$$\bar{V} \oplus \big(\bar{F} \otimes_F V_-'\big) \oplus (\bar{F} \otimes_F U) \cong \big(\bar{F} \otimes_F V_+'\big) \oplus (\bar{F} \otimes_F U).$$

The Jordan–Hölder Proposition 1.1.2 then implies that

$$\bar{V} \oplus \big(\bar{F} \otimes_F V_-'\big) \cong \bar{F} \otimes_F V_+'.$$

By repeating this argument we arrive after finitely many steps at an $F[G]$-module V such that

$$\bar{V} \cong \bar{F} \otimes_F V.$$

Step 2: Let now V be a simple $F[G]$-module, and let $\bar{F} \otimes_F V = \bar{V}_1 \oplus \cdots \oplus \bar{V}_m$ be a decomposition into simple $\bar{F}[G]$-modules. By Step 1 we find $F[G]$-modules V_i such that

$$\bar{V}_i \cong \bar{F} \otimes_F V_i.$$

For any $1 \leq i \leq m$, the $F[G]$-module V_i necessarily is simple, and we have $\mathrm{Hom}_{\bar{F}[G]}(\bar{F} \otimes_F V_i, \bar{F} \otimes_F V) \neq \{0\}$. Hence $\mathrm{Hom}_{F[G]}(V_i, V) \neq \{0\}$ by Lemma 2.7.1. It follows that $V_i \cong V$ for any $1 \leq i \leq m$. By comparing dimensions we conclude that $m = 1$. This means that $\bar{F} \otimes_F V$ is a simple $\bar{F}[G]$-module. Using Lemma 2.7.1 again we obtain

$$\bar{F} \otimes_F \mathrm{End}_{F[G]}(V) = \mathrm{End}_{\bar{F}[G]}(\bar{F} \otimes_F V) = \bar{F}.$$

Hence $\mathrm{End}_{F[G]}(V) = F$ must be one-dimensional. This shows that F is a splitting field for G. *Step 3:* Let $H \subseteq G$ be any subgroup with exponent e_H. Since e_H divides e the field F *a fortiori* contains a primitive e_Hth root of unity. Hence F also is a splitting field for H. □

2.8 Properties of the Cartan–Brauer Triangle

We go back to the setting from the beginning of this chapter: k is an algebraically closed field of characteristic $p > 0$, R is a $(0, p)$-ring for k with maximal ideal $\mathfrak{m}_R = R\pi_R$, and K denotes the field of fractions of R.

The ring R will be called *splitting* for our finite group G if K contains a primitive eth root of unity ζ where e is the exponent of G. By Theorem 2.7.2 the field K, in this case, is a splitting field for any subgroup of G. This additional condition can easily be achieved by defining $K' := K(\zeta)$ and $R' := \{a \in K' : \mathrm{Norm}_{K'/K}(a) \in R\}$; then R' is a $(0, p)$-ring for k which is splitting for G.

It is our goal in this section to establish the deeper properties of the Cartan–Brauer triangle

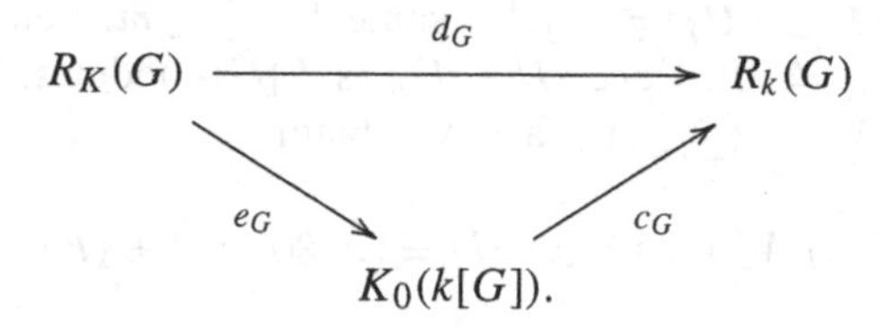

Lemma 2.8.1 *For any subgroup $H \subseteq G$ the diagram*

$$\begin{array}{ccc} R_K(H) & \xrightarrow{d_H} & R_k(H) \\ {\scriptstyle \mathrm{ind}_H^G} \downarrow & & \downarrow {\scriptstyle \mathrm{ind}_H^G} \\ R_K(G) & \xrightarrow{d_G} & R_k(G) \end{array}$$

is commutative.

Proof Let W be a finitely generated $K[H]$-module. We choose a lattice $L \subseteq W$ which is H-invariant. Then

$$\mathrm{ind}_H^G\big(d_H([W])\big) = \mathrm{ind}_H^G\big([L/\pi_R L]\big) = \big[k[G] \otimes_{k[H]} (L/\pi_R L)\big].$$

Moreover, $R[G] \otimes_{R[H]} L \cong L^{[G:H]}$ is a G-invariant lattice in $K[G] \otimes_{K[H]} W \cong W^{[G:H]}$ (compare the proof of Remark 2.3.2). Hence

$$\begin{aligned} d_G(\operatorname{ind}_H^G([W])) &= d_G([K[G] \otimes_{K[H]} W]) \\ &= [(R[G] \otimes_{R[H]} L)/\pi_R(R[G] \otimes_{R[H]} L)] \\ &= [k[G] \otimes_{k[H]} (L/\pi_R L)]. \end{aligned}$$

□

Lemma 2.8.2 *We have*

$$R_k(G) = \sum_{H \in \mathcal{H}_e} \operatorname{ind}_H^G(R_k(H)).$$

Proof Since the $\operatorname{ind}_H^G(R_k(H))$ are ideals in $R_k(G)$ it suffices to show that the unit element $1_{k[G]} \in R_k(G)$ lies in the right-hand side. We choose R to be splitting for G. By Brauer's induction Theorem 2.6.1 we have

$$1_{K[G]} \in \sum_{H \in \mathcal{H}_e} \operatorname{ind}_H^G(R_K(H))$$

where $1_{K[G]}$ is the unit element in $R_K(G)$. Using Lemma 2.8.1 we obtain

$$\begin{aligned} d_G(1_{K[G]}) &\in \sum_{H \in \mathcal{H}_e} d_G(\operatorname{ind}_H^G(R_K(H))) = \sum_{H \in \mathcal{H}_e} \operatorname{ind}_H^G(d_H(R_K(H))) \\ &\subseteq \sum_{H \in \mathcal{H}_e} \operatorname{ind}_H^G(R_k(H)). \end{aligned}$$

It is trivial to see that $d_G(1_{K[G]}) = 1_{k[G]}$. □

Theorem 2.8.3 *The decomposition homomorphism*

$$d_G: \quad R_K(G) \longrightarrow R_k(G)$$

is surjective.

Proof By Lemma 2.8.1 we have

$$d_G(R_K(G)) \supseteq \sum_{H \in \mathcal{H}_e} d_G(\operatorname{ind}_H^G(R_K(H))) = \sum_{H \in \mathcal{H}_e} \operatorname{ind}_H^G(d_H(R_K(H))).$$

Because of Lemma 2.8.2 it therefore suffices to show that $d_H(R_K(H)) = R_k(H)$ for any $H \in \mathcal{H}_e$. This means we are reduced to proving our assertion in the case where the group G is elementary. Then $G = H \times P$ is the direct product of a group H of order prime to p and a p-group P. By Proposition 1.7.1 it suffices to show that the class $[W] \in R_k(G)$, for any simple $k[G]$-module W, lies in the image of d_G. Viewed as a $k[P]$-module W must contain the trivial $k[P]$-module by Proposition 2.2.7. We deduce that

$$W^P := \{w \in W : gw = w \text{ for any } g \in P\} \neq \{0\}.$$

Since P is a normal subgroup of G the $k[P]$-submodule W^P in fact is a $k[G]$-submodule of W. But W is simple. Hence $W^P = W$ which means that $k[G]$ acts on W through the projection map $k[G] \longrightarrow k[H]$. According to Corollary 2.2.6 we find a simple $K[H]$-module V together with a G-invariant lattice $L \subseteq V$ such that $L/\pi_R L \cong W$ as $k[H]$-modules. Viewing V as a $K[G]$-module through the projection map $K[G] \longrightarrow K[H]$ we obtain $[V] \in R_K(G)$ and $d_G([V]) = [W]$. □

Theorem 2.8.4 *Let p^m be the largest power of p which divides the order of G; the Cartan homomorphism $c_G : K_0(k[G]) \longrightarrow R_k(G)$ is injective, its cokernel is finite, and $p^m R_k(G) \subseteq \mathrm{im}(c_G)$.*

Proof Step 1: We show that $p^m R_k(G) \subseteq \mathrm{im}(c_G)$ holds true. It is trivial from the definition of the Cartan homomorphism that, for any subgroup $H \subseteq G$, the diagram

$$\begin{array}{ccc} K_0(k[H]) & \xrightarrow{c_H} & R_k(H) \\ {\scriptstyle [P] \longmapsto [k[G] \otimes_{k[H]} P]} \downarrow & & \downarrow {\scriptstyle \mathrm{ind}_H^G} \\ K_0(k[G]) & \xrightarrow{c_G} & R_k(G) \end{array}$$

is commutative. It follows that

$$\mathrm{ind}_H^G\big(\mathrm{im}(c_H)\big) \subseteq \mathrm{im}(c_G).$$

Lemma 2.8.2 therefore reduces us to the case that G is an elementary group. Let W be any simple $k[G]$-module. With the notations of the proof of Theorem 2.8.3 we have seen there that $k[G]$ acts on W through the projection map $k[G] \longrightarrow k[H]$. Viewed as a $k[H]$-module W is projective by Remark 1.7.3. Hence $k[G] \otimes_{k[H]} W$ is a finitely generated projective $k[G]$-module. We claim that

$$c_G\big([k[G] \otimes_{k[H]} W]\big) = |G/H| \cdot [W]$$

holds true. Using the above commutative diagram as well as Proposition 2.3.4 we obtain

$$c_G\big([k[G] \otimes_{k[H]} W]\big) = \mathrm{ind}_H^G([W]) = \big[k[G] \otimes_{k[H]} k\big] \cdot [W].$$

In order to analyze the $k[G]$-module $k[G] \otimes_{k[H]} k$ let $h, h' \in H$, $g \in P$, and $a \in k$. Then

$$h\big(gh' \otimes a\big) = ghh' \otimes a = g \otimes a = gh' \otimes a.$$

This shows that H acts trivially on $k[G] \otimes_{k[H]} k$. In other words, $k[G]$ acts on $k[G] \otimes_{k[H]} k$ through the projection map $k[G] \longrightarrow k[P]$. It then follows from Proposition 2.2.7 that all simple subquotients in a composition series of the $k[G]$-

module $k[G] \otimes_{k[H]} k$ are trivial $k[G]$-modules. We conclude that

$$\big[k[G] \otimes_{k[H]} k\big] = \dim_k\big(k[G] \otimes_{k[H]} k\big) \cdot 1 = |G/H| \cdot 1$$

(where $1 \in R_k(G)$ is the unit element).

Step 2: We know from Proposition 1.7.1 that, as an abelian group, $R_k(G) \cong \mathbb{Z}^r$ for some $r \geq 1$. It therefore follows from Step 1 that $R_k(G)/\operatorname{im}(c_G)$ is isomorphic to a factor group of the finite group $\mathbb{Z}^r/p^m\mathbb{Z}^r$.

Step 3: It is a consequence of Proposition 1.7.4 that $K_0(k[G])$ and $R_k(G)$ are isomorphic to $\mathbb{Z}^r$ for the same integer $r \geq 1$. Hence

$$\mathrm{id} \otimes c_G\colon \quad \mathbb{Q} \otimes_{\mathbb{Z}} K_0\big(k[G]\big) \longrightarrow \mathbb{Q} \otimes_{\mathbb{Z}} R_k(G)$$

is a linear map between two $\mathbb{Q}$-vector spaces of the same finite dimension r. Its injectivity is equivalent to its surjectivity. Let $a \in \mathbb{Q}$ and $x \in R_k(G)$. By Step 1 we find an element $y \in K_0(k[G])$ such that $c_G(y) = p^m x$. Then

$$(\mathrm{id} \otimes c_G)\left(\frac{a}{p^m} \otimes y\right) = \frac{a}{p^m} \otimes p^m x = a \otimes x.$$

This shows that $\mathrm{id} \otimes c_G$ and consequently c_G are injective. □

In order to discuss the third homomorphism e_G we first introduce two bilinear forms. We start from the maps

$$(\mathfrak{M}_{K[G]}/\cong) \times (\mathfrak{M}_{K[G]}/\cong) \longrightarrow \mathbb{Z}$$
$$\big(\{V\}, \{W\}\big) \longmapsto \dim_K \mathrm{Hom}_{K[G]}(V, W)$$

and

$$(\mathcal{M}_{k[G]}/\cong) \times (\mathfrak{M}_{k[G]}/\cong) \longrightarrow \mathbb{Z}$$
$$\big(\{P\}, \{V\}\big) \longmapsto \dim_k \mathrm{Hom}_{k[G]}(P, V).$$

They extend to $\mathbb{Z}$-bilinear maps

$$\mathbb{Z}[\mathfrak{M}_{K[G]}] \times \mathbb{Z}[\mathfrak{M}_{K[G]}] \longrightarrow \mathbb{Z}$$

and

$$\mathbb{Z}[\mathcal{M}_{k[G]}] \times \mathbb{Z}[\mathfrak{M}_{k[G]}] \longrightarrow \mathbb{Z}.$$

Since $K[G]$ is semisimple we have, for any exact sequence $0 \to V_1 \to V \to V_2 \to 0$ in $\mathfrak{M}_{K[G]}$, that $V \cong V_1 \oplus V_2$ and hence that

$$\dim_K \mathrm{Hom}_{K[G]}(V, W) - \dim_K \mathrm{Hom}_{K[G]}(V_1, W) - \dim_K \mathrm{Hom}_{K[G]}(V_2, W) = 0.$$

The corresponding fact in the "variable" W holds as well, of course. The first map therefore induces a well-defined $\mathbb{Z}$-bilinear form

$$\begin{aligned} \langle\, , \rangle_{K[G]}\colon\quad R_K(G) \times R_K(G) &\longrightarrow \mathbb{Z} \\ \big([V],[W]\big) &\longmapsto \dim_K \mathrm{Hom}_{K[G]}(V,W). \end{aligned}$$

Even though $k[G]$ might not be semisimple any exact sequence $0 \to P_1 \to P \to P_2 \to 0$ in $\mathcal{M}_{k[G]}$ still satisfies $P \cong P_1 \oplus P_2$ as a consequence of Lemma 1.6.2.ii. Hence we again have

$$\dim_k \mathrm{Hom}_{k[G]}(P,V) - \dim_k \mathrm{Hom}_{k[G]}(P_1,V) - \dim_k \mathrm{Hom}_{k[G]}(P_2,V) = 0$$

for any V in $\mathfrak{M}_{k[G]}$. Furthermore, for any P in $\mathcal{M}_{k[G]}$ and any exact sequence $0 \to V_1 \to V \to V_2 \to 0$ in $\mathfrak{M}_{k[G]}$ we have, by the definition of projective modules, the exact sequence

$$0 \longrightarrow \mathrm{Hom}_{k[G]}(P,V_1) \longrightarrow \mathrm{Hom}_{k[G]}(P,V) \longrightarrow \mathrm{Hom}_{k[G]}(P,V_2) \longrightarrow 0.$$

Hence once more

$$\dim_k \mathrm{Hom}_{k[G]}(P,V) - \dim_k \mathrm{Hom}_{k[G]}(P,V_1) - \dim_k \mathrm{Hom}_{k[G]}(P,V_2) = 0.$$

This shows that the second map also induces a well-defined $\mathbb{Z}$-bilinear form

$$\begin{aligned} \langle\, , \rangle_{k[G]}\colon\quad K_0\big(k[G]\big) \times R_k(G) &\longrightarrow \mathbb{Z} \\ \big([P],[V]\big) &\longmapsto \dim_k \mathrm{Hom}_{k[G]}(P,V). \end{aligned}$$

If $\{V_1\},\ldots,\{V_r\}$ are the isomorphism classes of the simple $K[G]$-modules then $[V_1],\ldots,[V_r]$ is a $\mathbb{Z}$-basis of $R_K(G)$ by Proposition 1.7.1. We have

$$\big\langle [V_i],[V_j]\big\rangle_{K[G]} = \begin{cases} \dim_K \mathrm{End}_{K[G]}(V_i) & \text{if } i=j, \\ 0 & \text{if } i \neq j. \end{cases}$$

In particular, if K is a splitting field for G then

$$\big\langle [V_i],[V_j]\big\rangle_{K[G]} = \begin{cases} 1 & \text{if } i=j, \\ 0 & \text{if } i \neq j. \end{cases}$$

Let $\{P_1\},\ldots,\{P_t\}$ be the isomorphism classes of finitely generated indecomposable projective $k[G]$-modules. By Proposition 1.7.4.iii the $[P_1],\ldots,[P_t]$ form a $\mathbb{Z}$-basis of $K_0(k[G])$, and the $[P_1/\mathrm{Jac}(k[G])P_1],\ldots,[P_t/\mathrm{Jac}(k[G])P_t]$ form a $\mathbb{Z}$-basis of $R_k(G)$ by Proposition 1.7.4.i. We have

$$\begin{aligned}\mathrm{Hom}_{k[G]}\big(P_i, P_j/\mathrm{Jac}\big(k[G]\big)P_j\big) &= \mathrm{Hom}_{k[G]}\big(P_i/\mathrm{Jac}\big(k[G]\big)P_i, P_j/\mathrm{Jac}\big(k[G]\big)P_j\big)\\ &= \begin{cases}\mathrm{End}_{k[G]}(P_i/\mathrm{Jac}(k[G])P_i) & \text{if } i = j,\\ \{0\} & \text{if } i \neq j\end{cases}\\ &= \begin{cases}k & \text{if } i = j,\\ \{0\} & \text{if } i \neq j,\end{cases}\end{aligned}$$

where the latter identity comes from the fact that the algebraically closed field k is a splitting field for G. Hence

$$\big\langle [P_i], \big[P_j/\mathrm{Jac}\big(k[G]\big)P_j\big]\big\rangle_{k[G]} = \begin{cases}1 & \text{if } i = j,\\ 0 & \text{if } i \neq j.\end{cases}$$

Exercise 2.8.5

i. If K is a splitting field for G then the map

$$\begin{aligned} R_K(G) &\xrightarrow{\cong} \mathrm{Hom}_{\mathbb{Z}}\big(R_K(G), \mathbb{Z}\big)\\ x &\longmapsto \langle x, .\rangle_{K[G]}\end{aligned}$$

is an isomorphism of abelian groups.

ii. The maps

$$\begin{aligned} K_0\big(k[G]\big) &\xrightarrow{\cong} \mathrm{Hom}_{\mathbb{Z}}\big(R_k(G), \mathbb{Z}\big) \quad \text{and} \quad & R_k(G) &\xrightarrow{\cong} \mathrm{Hom}_{\mathbb{Z}}\big(K_0\big(k[G]\big), \mathbb{Z}\big)\\ y &\longmapsto \langle y, .\rangle_{k[G]} & z &\longmapsto \langle . , z\rangle_{k[G]}\end{aligned}$$

are isomorphisms of abelian groups.

Lemma 2.8.6 *We have*

$$\big\langle y, d_G(x)\big\rangle_{k[G]} = \big\langle e_G(y), x\big\rangle_{K[G]}$$

for any $y \in K_0(k[G])$ *and* $x \in R_K(G)$.

Proof It suffices to consider elements of the form $y = [P/\pi_R P]$ for some finitely generated projective $R[G]$-module P (see Proposition 2.1.1) and $x = [V]$ for some finitely generated $K[G]$-module V. We pick a G-invariant lattice $L \subseteq V$. The asserted identity then reads

$$\dim_k \mathrm{Hom}_{k[G]}(P/\pi_R P, L/\pi_R L) = \dim_K \mathrm{Hom}_{K[G]}(K \otimes_R P, V).$$

We have

$$\begin{aligned}\mathrm{Hom}_{k[G]}(P/\pi_R P, L/\pi_R L) &= \mathrm{Hom}_{R[G]}(P, L/\pi_R L)\\ &= \mathrm{Hom}_{R[G]}(P, L)/\mathrm{Hom}_{R[G]}(P, \pi_R L)\\ &= \mathrm{Hom}_{R[G]}(P, L)/\pi_R \mathrm{Hom}_{R[G]}(P, L)\\ &= k \otimes_R \mathrm{Hom}_{R[G]}(P, L) \qquad (2.8.1)\end{aligned}$$

where the second identity comes from the projectivity of P as an $R[G]$-module. On the other hand

$$\begin{aligned}\mathrm{Hom}_{K[G]}(K \otimes_R P, V) &= \mathrm{Hom}_{R[G]}(P, V)\\ &= \mathrm{Hom}_{R[G]}\Big(P, \bigcup_{i\geq 0} \pi_R^{-i} L\Big)\\ &= \bigcup_{i\geq 0} \mathrm{Hom}_{R[G]}(P, \pi_R^{-i} L)\\ &= \bigcup_{i\geq 0} \pi_R^{-i} \mathrm{Hom}_{R[G]}(P, L)\\ &= K \otimes_R \mathrm{Hom}_{R[G]}(P, L). \qquad (2.8.2)\end{aligned}$$

For the third identity one has to observe that, since P is finitely generated as an R-module, any R-module homomorphism $P \longrightarrow V = \bigcup_{i\geq 0} \pi_R^{-i} L$ has to have its image inside $\pi_R^{-i} L$ for some sufficiently large i.

Both, P being a direct summand of some $R[G]^m$ (Remark 2.1.2) and L by definition are free R-modules. Hence $\mathrm{Hom}_R(P, L)$ is a finitely generated free R-module. The ring R being noetherian the R-submodule $\mathrm{Hom}_{R[G]}(P, L)$ is finitely generated as well. Lemma 2.2.1.i then implies that $\mathrm{Hom}_{R[G]}(P, L) \cong R^s$ is a free R-module. We now deduce from (2.8.1) and (2.8.2) that

$$\mathrm{Hom}_{k[G]}(P/\pi_R P, L/\pi_R L) \cong k^s \quad \text{and} \quad \mathrm{Hom}_{K[G]}(K \otimes_R P, V) \cong K^s,$$

respectively. □

Theorem 2.8.7 *The homomorphism* $e_G : K_0(k[G]) \longrightarrow R_K(G)$ *is injective and its image is a direct summand of* $R_K(G)$.

Proof Step 1: We assume that R is splitting for G. By Theorem 2.8.3 the map $d_G : R_K(G) \longrightarrow R_k(G)$ is surjective. Since, by Proposition 1.7.1, $R_k(G)$ is a free abelian group we find a homomorphism $s : R_k(G) \longrightarrow R_K(G)$ such that $d_G \circ s = \mathrm{id}$. It follows that

$$\mathrm{Hom}(s, \mathbb{Z}) \circ \mathrm{Hom}(d_G, \mathbb{Z}) = \mathrm{Hom}(d_G \circ s, \mathbb{Z}) = \mathrm{Hom}(\mathrm{id}, \mathbb{Z}) = \mathrm{id}.$$

Hence the map

$$\operatorname{Hom}(d_G, \mathbb{Z}) : \operatorname{Hom}_{\mathbb{Z}}(R_k(G), \mathbb{Z}) \longrightarrow \operatorname{Hom}_{\mathbb{Z}}(R_K(G), \mathbb{Z})$$

is injective and

$$\operatorname{Hom}_{\mathbb{Z}}(R_K(G), \mathbb{Z}) = \operatorname{im}(\operatorname{Hom}(d_G, \mathbb{Z})) \oplus \ker(\operatorname{Hom}(s, \mathbb{Z})).$$

But because of Lemma 2.8.6 the map $\operatorname{Hom}(d_G, \mathbb{Z})$ corresponds under the isomorphisms in Exercise 2.8.5 to the homomorphism e_G.

Step 2: For general R we use, as described at the beginning of this section a larger $(0, p)$-ring R' for k which contains R and is splitting for G. Let K' denote the field of fractions of R'. It P is a finitely generated projective $R[G]$-module then $R' \otimes_R P = R'[G] \otimes_{R[G]} P$ is a finitely generated projective $R'[G]$-module such that

$$\begin{aligned}(R' \otimes_R P)/\pi_{R'}(R' \otimes_R P) &= (R'/\pi_{R'}R') \otimes_R P = (R/\pi_R R) \otimes_R P \\ &= P/\pi_R P\end{aligned}$$

(recall that $k = R/\pi_R R = R'/\pi_{R'}R'$). This shows that the diagram

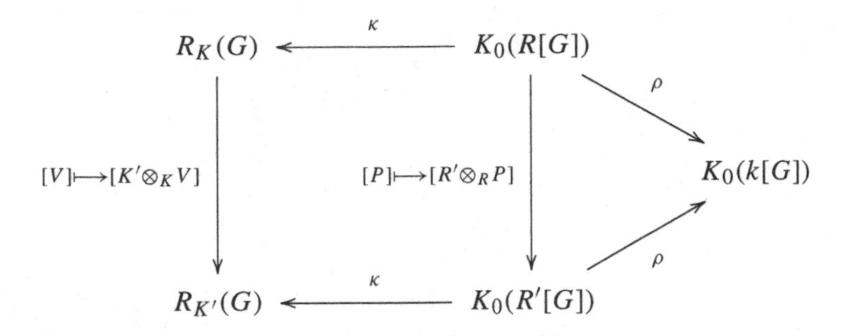

is commutative. Hence

$$\begin{array}{ccccc} K_0(k[G]) & \xrightarrow{e_G} & R_K(G) & \xrightarrow{\operatorname{Tr}} & \operatorname{Cl}(G, K) \\ & \searrow^{e_G} & \downarrow{\scriptstyle [V]\mapsto[K'\otimes_K V]} & & \downarrow{\scriptstyle \subseteq} \\ & & R_{K'}(G) & \xrightarrow{\operatorname{Tr}} & \operatorname{Cl}(G, K') \end{array}$$

is commutative. The oblique arrow is injective by Step 1 and so then is the upper left horizontal arrow. The two right horizontal arrows are injective by Corollary 2.3.7.i. This implies that the middle vertical arrow is injective and therefore induces an injective homomorphism

$$R_K(G)/\operatorname{im}(e_G) \longrightarrow R_{K'}(G)/\operatorname{im}(e_G).$$

By Step 1 the target $R_{K'}(G)/\operatorname{im}(e_G)$ is isomorphic to a direct summand of the free abelian group $R_{K'}(G)$. It follows that $R_K(G)/\operatorname{im}(e_G)$ is isomorphic to a subgroup in a finitely generated free abelian group and hence is a free abelian group by the elementary divisor theorem. We conclude that

$$R_K(G) \cong \operatorname{im}(e_G) \oplus R_K(G)/\operatorname{im}(e_G).$$ □

We choose R to be splitting for G, and we fix the $\mathbb{Z}$-bases of the three involved Grothendieck groups as described before Exercise 2.8.5. Let E, D, and C denote the matrices which describe the homomorphisms e_G, d_G, and c_G, respectively, with respect to these bases. We, of course, have

$$DE = C.$$

Lemma 2.8.6 says that D is the transpose of E. It follows that the quadratic Cartan matrix C of $k[G]$ is symmetric.

Chapter 3
The Brauer Character

As in the last chapter we fix an algebraically closed field k of characteristic $p > 0$, and we let G be a finite group. We also fix a $(0, p)$-ring R for k which is splitting for G. Let $\mathfrak{m}_R = R\pi_R$ denote the maximal ideal and K the field of fractions of R.

3.1 Definitions

As in the semisimple case we let $\mathrm{Cl}(G, k)$ denote the k-vector space of all k-valued class functions on G, i.e. functions on G which are constant on conjugacy classes. For any $V \in \mathfrak{M}_{k[G]}$ we introduce its *k-character*

$$\begin{aligned} \chi_V \colon \quad G &\longrightarrow k \\ g &\longmapsto \mathrm{tr}(g; V), \end{aligned}$$

which is a function in $\mathrm{Cl}(G, k)$. Let $0 \to V_1 \xrightarrow{\alpha} V \xrightarrow{\beta} V_2 \to 0$ be an exact sequence in $\mathfrak{M}_{k[G]}$. For $i = 1, 2$ we let A_i be the matrix of $V_i \xrightarrow{g} V_i$ with respect to some k-basis $e_1^{(i)}, \ldots, e_{d_i}^{(i)}$. We put $e_j := \alpha(e_j^{(1)})$ for $1 \le j \le d_1$ and we choose $e_j \in V$, for $d_1 < j \le d_1 + d_2$, such that $\beta(e_{j-d_1}) = e_j^{(2)}$. Then the matrix of $V \xrightarrow{g} V$ with respect to the basis $e_1, \ldots, e_{d_1}, \ldots, e_{d_1+d_2}$ is of the form

$$\begin{pmatrix} A_1 & A \\ 0 & A_2 \end{pmatrix}.$$

Since the trace is the sum of the diagonal entries of the respective matrix we see that

$$\chi_V(g) = \chi_{V_1}(g) + \chi_{V_2}(g) \quad \text{for any } g \in G.$$

It follows that

P. Schneider, *Modular Representation Theory of Finite Groups*,
DOI 10.1007/978-1-4471-4832-6_3, © Springer-Verlag London 2013

$$\begin{aligned} \mathrm{Tr}\colon \quad R_k(G) &\longrightarrow \mathrm{Cl}(G,k) \\ [V] &\longmapsto \chi_V \end{aligned}$$

is a well-defined homomorphism.

Proposition 3.1.1 *The k-characters $\chi_V \in \mathrm{Cl}(G,k)$, for $\{V\} \in \widehat{k[G]}$, are k-linearly independent.*

Proof Let $\widehat{k[G]} = \{\{V_1\},\ldots,\{V_r\}\} = \hat{A}$ where $A := k[G]/\mathrm{Jac}(k[G])$. Since $k[G]$ is artinian A is semisimple (cf. Proposition 1.1.2.vi). By Wedderburn's structure theory of semisimple k-algebras we have

$$A = \prod_{i=1}^{r} \mathrm{End}_{D_i}(V_i)$$

where $D_i := \mathrm{End}_{k[G]}(V_i)$. Since k is algebraically closed we, in fact, have $D_i = k$. For any $1 \le i \le r$ we pick an element $\alpha_i \in \mathrm{End}_k(V_i)$ with $\mathrm{tr}(\alpha_i) = 1$. By the above product decomposition of A we then find elements $a_i \in k[G]$, for $1 \le i \le r$, such that

$$(V_j \xrightarrow{a_i} V_j) = \begin{cases} \alpha_i & \text{if } i = j, \\ 0 & \text{if } i \ne j. \end{cases}$$

If we extend each χ_{V_j} by k-linearity to a k-linear form $\widetilde{\chi}_{V_j} : k[G] \longrightarrow k$ then

$$\widetilde{\chi}_{V_j}(a_i) = \mathrm{tr}(a_i; V_j) = \begin{cases} 1 & \text{if } i = j, \\ 0 & \text{if } i \ne j. \end{cases}$$

We now suppose that $\sum_{j=1}^{r} c_j \chi_{V_j} = 0$ in $\mathrm{Cl}(G,k)$ with $c_j \in k$. Then also $\sum_{j=1}^{r} c_j \widetilde{\chi}_{V_j} = 0$ in $\mathrm{Hom}_k(k[G],k)$ and hence $0 = \sum_{j=1}^{r} c_j \widetilde{\chi}_{V_j}(a_i) = c_i$ for any $1 \le i \le r$. □

Since, by Proposition 1.7.1, the vectors $1 \otimes [V]$, for $\{V\} \in \widehat{k[G]}$, form a basis of the k-vector space $k \otimes_{\mathbb{Z}} R_k(G)$ it follows that

$$\begin{aligned} \mathrm{Tr}\colon \quad k \otimes_{\mathbb{Z}} R_k(G) &\longrightarrow \mathrm{Cl}(G,k) \\ c \otimes [V] &\longrightarrow c\chi_V \end{aligned}$$

is an injective k-linear map. But the canonical map $R_k(G) \longrightarrow k \otimes_{\mathbb{Z}} R_k(G)$ is not injective. Therefore the k-character χ_V does not determine the isomorphism class $\{V\}$. This is the reason that in the present setting the k-characters are of limited use.

Definition An element $g \in G$ is called p-regular, resp. p-unipotent, if the order of g is prime to p, resp. is a power of p.

Lemma 3.1.2 *For any $g \in G$ there exist uniquely determined elements g_{reg} and g_{uni} in G such that*

– g_{reg} *is* p*-regular,* g_{uni} *is* p*-unipotent, and*
– $g = g_{\mathrm{reg}} g_{\mathrm{uni}} = g_{\mathrm{uni}} g_{\mathrm{reg}}$;

moreover, g_{reg} *and* g_{uni} *are powers of* g.

Proof Let $p^s m$ with $p \nmid m$ be the order of g. We choose integers a and b such that $ap^s + bm = 1$, and we define $g_{\mathrm{reg}} := g^{ap^s}$ and $g_{\mathrm{uni}} := g^{bm}$. Then

$$g = g^{ap^s + bm} = g_{\mathrm{reg}} g_{\mathrm{uni}} = g_{\mathrm{uni}} g_{\mathrm{reg}}.$$

Furthermore, $g_{\mathrm{reg}}^m = g^{ap^s m} = 1$ and $g_{\mathrm{uni}}^{p^s} = g^{bmp^s} = 1$; hence g_{reg} is p-regular and g_{uni} is p-unipotent. Let $g = g_r g_u = g_u g_r$ be another decomposition with p-regular g_r and p-unipotent g_u. One checks that g_r and g_u have the order m and p^s, respectively. Then $g_{\mathrm{reg}} = g^{ap^s} = g^{1-bm} = g_r^{1-bm} g_u^{ap^s} = g_r$ and hence also $g_{\mathrm{uni}} = g_u$. □

Let e denote the exponent of G and let $e = e' p^s$ with $p \nmid e'$. We consider any finitely generated $k[G]$-module V and any element $g \in G$. Let $\zeta_1(g, V), \ldots, \zeta_d(g, V)$, where $d := \dim_k V$, be all eigenvalues of the k-linear endomorphism $V \xrightarrow{g} V$. We list any eigenvalue as many times as its multiplicity as a zero of the characteristic polynomial of $V \xrightarrow{g} V$ prescribes. We have

$$\chi_V(g) = \zeta_1(g, V) + \cdots + \zeta_d(g, V).$$

The following are easy facts:

1. The sequence $(\zeta_1(g, V), \ldots, \zeta_d(g, V))$ depends up to its ordering only on the isomorphism class $\{V\}$.
2. If ζ is an eigenvalue of $V \xrightarrow{g} V$ then ζ^j, for any $j \geq 0$, is an eigenvalue of $V \xrightarrow{g^j} V$. It follows that $\zeta_i(g, V)^{\mathrm{order}(g)} = 1$ for any $1 \leq i \leq d$. In particular, each $\zeta_i(g, V)$ is an eth root of unity. Is g p-regular then the $\zeta_i(g, V)$ are e'th roots of unity.
3. If $0 \to V_1 \to V \to V_2 \to 0$ is an exact sequence in $\mathfrak{M}_{k[G]}$ then the sequence $(\zeta_1(g, V), \ldots, \zeta_d(g, V))$ is, up to a reordering, the union of the two sequences $(\zeta_1(g, V_1), \ldots, \zeta_{d_1}(g, V_1))$ and $(\zeta_1(g, V_2), \ldots, \zeta_{d_2}(g, V_2))$ (where $d_i := \dim_k V_i$).

Lemma 3.1.3 *For any finitely generated* $k[G]$*-module* V *and any* $g \in G$ *the sequences* $(\zeta_1(g, V), \ldots, \zeta_d(g, V))$ *and* $(\zeta_1(g_{\mathrm{reg}}, V), \ldots, \zeta_d(g_{\mathrm{reg}}, V))$ *coincide up to a reordering; in particular, we have*

$$\chi_V(g) = \chi_V(g_{\mathrm{reg}}).$$

Proof Since the order of g_{reg} is prime to p the vector space

$$V = V_1 \oplus \cdots \oplus V_t$$

decomposes into the different eigenspaces V_j for the linear endomorphism $V \xrightarrow{g_{\text{reg}}} V$. The elements $g_{\text{reg}} g_{\text{uni}} = g_{\text{uni}} g_{\text{reg}}$ commute. Hence g_{uni} respects the eigenspaces of g_{reg}, i.e. $g_{\text{uni}}(V_j) = V_j$ for any $1 \le j \le t$. The cyclic group $\langle g_{\text{uni}} \rangle$ is a p-group. By Proposition 2.2.7 the only simple $k[\langle g_{\text{uni}} \rangle]$-module is the trivial module. This implies that 1 is the only eigenvalue of $V \xrightarrow{g_{\text{uni}}} V$. We therefore find a basis of V_j with respect to which the matrix of $g_{\text{uni}}|V_j$ is of the form $\left(\begin{smallmatrix} 1 & & * \\ & \ddots & \\ 0 & & 1 \end{smallmatrix}\right)$. The matrix of $g_{\text{reg}}|V_j$ is the diagonal matrix $\left(\begin{smallmatrix} \zeta_j & & 0 \\ & \ddots & \\ 0 & & \zeta_j \end{smallmatrix}\right)$ where ζ_j is the corresponding eigenvalue. The matrix of $g|V_j$ then is $\left(\begin{smallmatrix} \zeta_j & & * \\ & \ddots & \\ 0 & & \zeta_j \end{smallmatrix}\right)$. It follows that $g|V_j$ has a single eigenvalue which coincides with the eigenvalue of $g_{\text{reg}}|V_j$. □

The subset

$$G_{\text{reg}} := \{g \in G : g \text{ is } p\text{-regular}\}$$

consists of full conjugacy classes of G. We therefore may introduce the k-vector space $\mathrm{Cl}(G_{\text{reg}}, k)$ of k-valued class functions on G_{reg}. There is the obvious map $\mathrm{Cl}(G, k) \longrightarrow \mathrm{Cl}(G_{\text{reg}}, k)$ which sends a function on G to its restriction to G_{reg}. Proposition 3.1.1 and Lemma 3.1.3 together imply that the composed map

$$\mathrm{Tr}_{\text{reg}}: \quad k \otimes_{\mathbb{Z}} R_k(G) \xrightarrow{\mathrm{Tr}} \mathrm{Cl}(G, k) \longrightarrow \mathrm{Cl}(G_{\text{reg}}, k)$$

still is injective. We will see later on that Tr_{reg} in fact is an isomorphism.

Remark 3.1.4 Let $\xi \in K^\times$ be any root of unity; then $\xi \in R^\times$.

Proof We have $\xi = a\pi_R^j$ with $a \in R^\times$ and $j \in \mathbb{Z}$. If $\xi^m = 1$ with $m \ge 1$ then $1 = a^m \pi_R^{jm}$ which implies $\pi_R^{jm} \in R^\times$. It follows that $j = 0$ and consequently that $\xi = a \in R^\times$. □

Let $\mu_{e'}(K)$ and $\mu_{e'}(k)$ denote the subgroup of $K^\times$ and $k^\times$, respectively, of all e'th roots of unity. Both groups are cyclic of order e', $\mu_{e'}(K)$ since R is splitting for G, and $\mu_{e'}(k)$ since k is algebraically closed of characteristic prime to e'. Since $\mu_{e'}(K) \subseteq R^\times$ by Remark 3.1.4 the homomorphism

$$\begin{aligned} \mu_{e'}(K) &\longrightarrow \mu_{e'}(k) \\ \xi &\longmapsto \xi + \mathfrak{m}_R \end{aligned}$$

is well-defined.

Lemma 3.1.5 *The map* $\mu_{e'}(K) \xrightarrow{\cong} \mu_{e'}(k)$ *is an isomorphism.*

Proof The elements in $\mu_{e'}(K)$ are precisely the roots of the polynomial $X^{e'} - 1 \in K[X]$. If $\xi_1 \ne \xi_2$ in $\mu_{e'}(K)$ were mapped to the same element in $\mu_{e'}(k)$ then the same polynomial $X^{e'}-1$ but viewed in $k[X]$ would have a zero of multiplicity >1. But this polynomial is separable since $p \nmid e'$. Hence the map is injective and then also bijective. □

We denote the inverse of the isomorphism in Lemma 3.1.5 by

$$\begin{aligned} \mu_{e'}(k) &\longrightarrow \mu_{e'}(K) \subseteq R \\ \xi &\longmapsto [\xi]. \end{aligned}$$

The element $[\xi] \in R$ is called the *Teichmüller representative* of ξ.

The isomorphism in Lemma 3.1.5 allows us to introduce, for any finitely generated $k[G]$-module V (with $d := \dim_k V$), the K-valued class function

$$\begin{aligned} \beta_V: \quad G_{\mathrm{reg}} &\longrightarrow K \\ g &\longmapsto [\zeta_1(g,V)] + \cdots + [\zeta_d(g,V)] \end{aligned}$$

on G_{reg}. It is called the *Brauer character* of V. By construction we have

$$\beta_V(g) \equiv \chi_V(g) \bmod \mathfrak{m}_R \quad \text{for any } g \in G_{\mathrm{reg}}.$$

The first fact in the list before Lemma 3.1.3 implies that β_V only depends on the isomorphism class $\{V\}$, whereas the last fact implies that

$$\beta_V = \beta_{V_1} + \beta_{V_2}$$

for any exact sequence $0 \to V_1 \to V \to V_2 \to 0$ in $\mathfrak{M}_{k[G]}$. Therefore, if we let $\mathrm{Cl}(G_{\mathrm{reg}}, K)$ denote the K-vector space of all K-valued class functions on G_{reg} then

$$\begin{aligned} \mathrm{Tr}_B: \quad R_k(G) &\longrightarrow \mathrm{Cl}(G_{\mathrm{reg}}, K) \\ [V] &\longmapsto \beta_V \end{aligned}$$

is a well-defined homomorphism.

3.2 Properties

Lemma 3.2.1 *The Brauer characters $\beta_V \in \mathrm{Cl}(G_{\mathrm{reg}}, K)$, for $\{V\} \in \widehat{k[G]}$, are K-linearly independent.*

Proof Let $\sum_{\{V\} \in \widehat{k[G]}} c_V \beta_V = 0$ with $c_V \in K$. By multiplying the c_V by a high enough power of π_R we may assume that all c_V lie in R. Then

$$\sum_{\{V\}} (c_V \bmod \mathfrak{m}_R) \chi_V = 0,$$

and Proposition 3.1.1 implies that all c_V must lie in $\mathfrak{m}_R$. We therefore may write $c_V = \pi_R d_V$ with $d_V \in R$ and obtain

$$0 = \sum_{\{V\}} c_V \beta_V = \pi_R \sum_{\{V\}} d_V \beta_V, \quad \text{hence} \quad \sum_{\{V\}} d_V \beta_V = 0.$$

Applying Proposition 3.1.1 again gives $d_V \in \mathfrak{m}_R$ and $c_V \in \mathfrak{m}_V^2$. Proceeding inductively in this way we deduce that $c_V \in \bigcap_{i \geq 0} \mathfrak{m}_R^i = \{0\}$ for any $\{V\} \in \widehat{k[G]}$. □

Proposition 3.2.2 *The diagram*

$$\begin{array}{ccc} R_K(G) & \xrightarrow{d_G} & R_k(G) \\ \Big\downarrow{\scriptstyle \mathrm{Tr}} & & \Big\downarrow{\scriptstyle \mathrm{Tr}_B} \\ \mathrm{Cl}(G, K) & \xrightarrow{\mathrm{res}} & \mathrm{Cl}(G_{\mathrm{reg}}, K), \end{array}$$

where "res" denotes the map of restricting functions from G to G_{reg}, is commutative.

Proof Let V be any finitely generated $K[G]$-module. We choose a lattice $L \subseteq V$ which is G-invariant and put $W := L/\pi_R L$. Then $d_G([V]) = [W]$, and our assertion becomes the statement that

$$\beta_W(g) = \chi_V(g) \quad \text{for any } g \in G_{\mathrm{reg}}$$

holds true. Let $e_1, \ldots, e_d$ be a K-basis of V such that $L = Re_1 + \cdots + Re_d$. Then $\bar{e}_1 := e_1 + \pi_R L, \ldots, \bar{e}_d := e_d + \pi_R L$ is a k-basis of W. We fix $g \in G_{\mathrm{reg}}$, and we form the matrices A_g and $\overline{A}_g$ of the linear endomorphisms $V \xrightarrow{g} V$ and $W \xrightarrow{g} W$, respectively, with respect to these bases. Obviously $\overline{A}_g$ is obtained from A_g by reducing its entries modulo $\mathfrak{m}_R$. Let $\zeta_1(g, V), \ldots, \zeta_d(g, V) \in K$, resp. $\zeta_1(g, W), \ldots, \zeta_d(g, W) \in k$, be the zeros, counted with multiplicity, of the characteristic polynomial of A_g, resp. $\overline{A}_g$. We have

$$\chi_V(g) = \mathrm{tr}(A_g) = \sum_i \zeta_i(g, V) \quad \text{and} \quad \chi_W(g) = \mathrm{tr}(\overline{A}_g) = \sum_i \zeta_i(g, W).$$

Clearly, the set $\{\zeta_i(g, V)\}_i$ is mapped by reduction modulo $\mathfrak{m}_R$ to the set $\{\zeta_i(g, W)\}_i$. Lemma 3.1.5 then implies that, up to a reordering, we have

$$\zeta_i(g, V) = \big[\zeta_i(g, W)\big].$$

We conclude that

$$\chi_V(g) = \sum_i \zeta_i(g, V) = \sum_i \big[\zeta_i(g, W)\big] = \beta_W(g).$$

□

Corollary 3.2.3 *The homomorphism*

$$\begin{aligned} \mathrm{Tr}_B: \quad K \otimes_{\mathbb{Z}} R_k(G) &\xrightarrow{\cong} \mathrm{Cl}(G_{\mathrm{reg}}, K) \\ c \otimes [V] &\longmapsto c\beta_V \end{aligned}$$

is an isomorphism.

Proof From Proposition 3.2.2 we obtain the commutative diagram

$$\begin{array}{ccc} K \otimes_{\mathbb{Z}} R_K(G) & \xrightarrow{\mathrm{id} \otimes d_G} & K \otimes_{\mathbb{Z}} R_k(G) \\ \downarrow{\scriptstyle \mathrm{Tr}} & & \downarrow{\scriptstyle \mathrm{Tr}_B} \\ \mathrm{Cl}(G, K) & \xrightarrow{\mathrm{res}} & \mathrm{Cl}(G_{\mathrm{reg}}, K) \end{array}$$

of K-linear maps. The left vertical map is an isomorphism by Corollary 2.3.7.ii. The map "res" clearly is surjective: For example, $\tilde{\varphi}(g) := \varphi(g_{\mathrm{reg}})$ is a preimage of $\varphi \in \mathrm{Cl}(G_{\mathrm{reg}}, K)$. Hence Tr_B is surjective. Its injectivity follows from Lemma 3.2.1. □

Corollary 3.2.4 *The number of isomorphism classes of simple $k[G]$-modules coincides with the number of conjugacy classes of p-regular elements in G.*

Proof The two numbers whose equality is asserted are the dimensions of the two K-vector spaces in the isomorphism of Corollary 3.2.3. □

Corollary 3.2.5 *The homomorphism* $\mathrm{Tr}_{\mathrm{reg}} : k \otimes_{\mathbb{Z}} R_k(G) \xrightarrow{\cong} \mathrm{Cl}(G_{\mathrm{reg}}, k)$ *is an isomorphism.*

Proof We know already that this k-linear map is injective. But by Corollary 3.2.4 the two k-vector spaces have the same finite dimension. Hence the map is bijective. □

We also may deduce a more conceptual formula for Brauer characters.

Corollary 3.2.6 *For any finitely generated $k[G]$-module V and any $g \in G_{\mathrm{reg}}$ we have*

$$\beta_V(g) = \mathrm{Tr}\big(d_{\langle g \rangle}^{-1}([V])\big)(g).$$

Proof By construction the element $\beta_V(g)$ only depends on V as a $k[\langle g \rangle]$-module where $\langle g \rangle \subseteq G$ is the cyclic subgroup generated by g. Since the order of $\langle g \rangle$ is prime to p the decomposition homomorphism $d_{\langle g \rangle}$ is an isomorphism by Corollary 2.2.6. The assertion therefore follows from Proposition 3.2.2 applied to the group $\langle g \rangle$. □

Lemma 3.2.7 *Let M be a finitely generated projective $R[G]$-module and put $V := K \otimes_R M$; we then have*

$$\chi_V(g) = 0 \quad \text{for any } g \in G \setminus G_{\mathrm{reg}}.$$

Proof We fix an element $g \in G \setminus G_{\mathrm{reg}}$. It generates a cyclic subgroup $\langle g \rangle \subseteq G$ whose order is divisible by p. The trace of g on V only depends on M viewed as an $R[G]$-module. We have $R[G] \cong R[\langle g \rangle]^{[G:\langle g \rangle]}$. Hence M, being isomorphic to a direct summand of some free $R[G]$-module, also is isomorphic to a direct summand of a free $R[\langle g \rangle]$-module. We see that M also is projective as an $R[\langle g \rangle]$-module (cf. Proposition 1.6.4). This observation reduces us (by replacing G by $\langle g \rangle$) to the case of a cyclic group G generated by an element whose order is divisible by p.

We write $G = C \times P$ as the direct product of a cyclic group C of order prime to p and a cyclic p-group $P \neq \{1\}$. First of all we note that by repeating the above observation for the subgroup $P \subseteq G$ we obtain that M is projective also as an $R[P]$-module. Let $g = g_C g_P$ with $g_C \in C$ and $g_P \in P$. We decompose

$$V = V_1 \oplus \cdots \oplus V_r$$

into its isotypic components V_i as an $R[C]$-module. Since K is a splitting field for any subgroup of G the V_i are precisely the different eigenspaces of the linear endomorphism $V \xrightarrow{g_C} V$. Let $\zeta_i \in K$ be the eigenvalue of g_C on V_i. Each V_i, of course, is a $K[G]$-submodule of V. In particular, we have

$$\chi_V(g) = \sum_{i=1}^{r} \zeta_i \operatorname{tr}(g_P; V_i).$$

It therefore suffices to show that $\operatorname{tr}(g_P; V_i) = 0$ for any $1 \leq i \leq r$. The element

$$\varepsilon_i := \frac{1}{|C|} \sum_{j=1}^{r} \zeta_i^{-j} g_C^j \in K[C]$$

is the idempotent such that $V_i = \varepsilon_i V$. But since $p \nmid |G|$ and because of Remark 3.1.4 we see that $\varepsilon_i \in R[C]$. It follows that we have the decomposition as an $R[G]$-module

$$M = M_1 \oplus \cdots \oplus M_r \quad \text{with } M_i := M \cap V_i.$$

Each M_i, being a direct summand of the projective $R[P]$-module M, is a finitely generated projective $R[P]$-module, and $V_i = K \otimes_R M_i$. This further reduces us (by replacing M by M_i and g by g_P) to the case where G is a cyclic p-group with generator $g \neq 1$.

We know from Propositions 1.7.4 and 2.2.7 that in this situation the (up to isomorphism) only finitely generated indecomposable projective $R[G]$-module is $R[G]$. Hence

$$M \cong R[G]^m$$

for some $m \geq 0$, and

$$\mathrm{tr}(g; V) = m\,\mathrm{tr}\big(g; K[G]\big).$$

Since $g \neq 1$ we have $gh \neq h$ for any $h \in G$. Using the K-basis $\{h\}_{h\in G}$ of $K[G]$ we therefore see that the corresponding matrix of $K[G] \xrightarrow{g} K[G]$ has all diagonal entries equal to zero. Hence

$$\mathrm{tr}\big(g; K[G]\big) = 0.$$

□

The above lemma can be rephrased by saying that there is a unique homomorphism

$$\mathrm{Tr}_{\mathrm{proj}}: \quad K_0\big(k[G]\big) \longrightarrow \mathrm{Cl}(G_{\mathrm{reg}}, K)$$

such that the diagram

$$\begin{array}{ccc} K_0(k[G]) & \xrightarrow{e_G} & R_K(G) \\ \downarrow{\scriptstyle \mathrm{Tr}_{\mathrm{proj}}} & & \downarrow{\scriptstyle \mathrm{Tr}} \\ \mathrm{Cl}(G_{\mathrm{reg}}, K) & \xrightarrow{\mathrm{ext}_0} & \mathrm{Cl}(G, K), \end{array}$$

where "ext_0" denotes the homomorphism which extends a function on G_{reg} to G by setting it equal to zero on $G \setminus G_{\mathrm{reg}}$, is commutative.

Proposition 3.2.8

i. *The map* $\mathrm{Tr}_{\mathrm{proj}} : K \otimes_{\mathbb{Z}} K_0(k[G]) \xrightarrow{\cong} \mathrm{Cl}(G_{\mathrm{reg}}, K)$ *is an isomorphism.*
ii. $\mathrm{im}(K_0(k[G]) \xrightarrow{e_G} R_K(G)) = \{x \in R_K(G) : \mathrm{Tr}(x)|G \setminus G_{\mathrm{reg}} = 0\}$.

Proof i. We have the commutative diagram

$$\begin{array}{ccc} K \otimes_{\mathbb{Z}} K_0(k[G]) & \xrightarrow{\mathrm{id} \otimes e_G} & K \otimes_{\mathbb{Z}} R_K(G) \\ \downarrow{\scriptstyle \mathrm{Tr}_{\mathrm{proj}}} & & \downarrow{\scriptstyle \mathrm{Tr}} \\ \mathrm{Cl}(G_{\mathrm{reg}}, K) & \xrightarrow{\mathrm{ext}_0} & \mathrm{Cl}(G, K). \end{array}$$

The maps $\mathrm{id} \otimes e_G$ and Tr are injective by Theorem 2.8.7 and Corollary 2.3.7.ii, respectively. Hence $\mathrm{Tr}_{\mathrm{proj}}$ is injective. Using Proposition 1.7.4 and Corollary 3.2.3 we obtain

$$\dim_K K \otimes_{\mathbb{Z}} K_0\big(k[G]\big) = \dim_K K \otimes_{\mathbb{Z}} R_k(G) = \dim_K \mathrm{Cl}(G_{\mathrm{reg}}, K).$$

It follows that $\mathrm{Tr}_{\mathrm{proj}}$ is bijective.

ii. As a consequence of Lemma 3.2.7 the left-hand side $\mathrm{im}(e_G)$ of the asserted identity is contained in the right-hand side. According to Theorem 2.8.7 there exists a subgroup $Z \subseteq R_K(G)$ such that $R_K(G) = \mathrm{im}(e_G) \oplus Z$ and *a fortiori*

$$K \otimes_{\mathbb{Z}} R_K(G) = \mathrm{im}(\mathrm{id} \otimes e_G) \oplus (K \otimes_{\mathbb{Z}} Z).$$

Suppose that $z \in Z$ is such that $\mathrm{Tr}(z)|G \setminus G_{\mathrm{reg}} = 0$. By i. we then find an element $x \in \mathrm{im}(\mathrm{id} \otimes e_G)$ such that $\mathrm{Tr}(z) = \mathrm{Tr}(x)$. It follows that

$$z = x \in Z \cap \mathrm{im}(\mathrm{id} \otimes e_G) \subseteq (K \otimes_{\mathbb{Z}} Z) \cap \mathrm{im}(\mathrm{id} \otimes e_G) = \{0\}.$$

Hence $z = 0$, and we see that any $x \in R_K(G)$ such that $\mathrm{Tr}(x)|G \setminus G_{\mathrm{reg}} = 0$ must be contained in $\mathrm{im}(e_G)$. □

Chapter 4
Green's Theory of Indecomposable Modules

4.1 Relatively Projective Modules

Let $f : A \longrightarrow B$ be any ring homomorphism. Any B-module M may be considered, via restriction of scalars, as an A-module. Any homomorphism $\alpha : M \longrightarrow N$ of B-modules automatically is a homomorphism of A-modules as well.

Definition A B-module P is called relatively projective (with respect to f) if for any pair of B-module homomorphisms

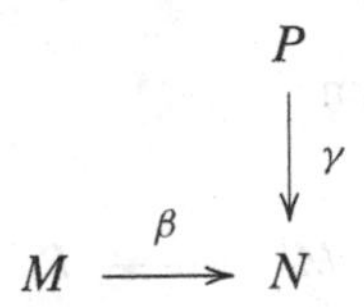

for which there is an A-module homomorphism $\alpha_0 : P \longrightarrow M$ such that $\beta \circ \alpha_0 = \gamma$ there also exists a B-module homomorphism $\alpha : P \longrightarrow M$ satisfying $\beta \circ \alpha = \gamma$.

We start with some very simple observations.

Remark

1. The existence of α_0 in the above definition implies that the image of β contains the image of γ. Hence we may replace N by $\mathrm{im}(\beta)$. This means that in the above definition it suffices to consider pairs (β, γ) with surjective β.
2. Any projective B-module is relatively projective.
3. If P is relatively projective and is projective as an A-module then P is a projective B-module.
4. Every B-module is relatively projective with respect to the identity homomorphism id_B.

Lemma 4.1.1 *For any B-module P the following conditions are equivalent:*

P. Schneider, *Modular Representation Theory of Finite Groups*,
DOI 10.1007/978-1-4471-4832-6_4, © Springer-Verlag London 2013

i. *P is relatively projective.*
ii. *For any (surjective) B-module homomorphism $\beta : M \longrightarrow P$ for which there is an A-module homomorphism $\sigma_0 : P \longrightarrow M$ such that $\beta \circ \sigma_0 = \mathrm{id}_P$ there also exists a B-module homomorphism $\sigma : P \longrightarrow M$ satisfying $\beta \circ \sigma = \mathrm{id}_P$.*

Proof i. $\Longrightarrow$ ii. We apply the definition to the pair

$$\begin{array}{ccc} & & P \\ & & \downarrow \mathrm{id}_P \\ M & \xrightarrow{\beta} & P. \end{array}$$

ii. $\Longrightarrow$ i. Let

$$\begin{array}{ccc} & & P \\ & & \downarrow \gamma \\ M & \xrightarrow{\beta} & N \end{array}$$

be any pair of B-module homomorphisms such that there is an A-module homomorphism $\alpha_0 : P \longrightarrow M$ with $\beta \circ \alpha_0 = \gamma$. By introducing the B-module

$$M' := \{(x, y) \in M \oplus P : \beta(x) = \gamma(y)\}$$

we obtain the commutative diagram

$$\begin{array}{ccc} M' & \xrightarrow{\mathrm{pr}_2} & P \\ \mathrm{pr}_1 \downarrow & & \downarrow \gamma \\ M & \xrightarrow{\beta} & N \end{array}$$

where $\mathrm{pr}_1((x, y)) := x$ and $\mathrm{pr}_2((x, y)) := y$. We observe that

$$\begin{aligned} \sigma_0 \colon \quad P &\longrightarrow M' \\ y &\longmapsto (\alpha_0(y), y) \end{aligned}$$

is a well-defined A-module homomorphism such that $\mathrm{pr}_2 \circ \sigma_0 = \mathrm{id}_P$. There exists therefore, by assumption, a B-module homomorphism $\sigma : P \longrightarrow M'$ satisfying $\mathrm{pr}_2 \circ \sigma = \mathrm{id}_P$. It follows that the B-module homomorphism $\alpha := \mathrm{pr}_1 \circ \sigma$ satisfies

$$\beta \circ \alpha = (\beta \circ \mathrm{pr}_1) \circ \sigma = (\gamma \circ \mathrm{pr}_2) \circ \sigma = \gamma \circ (\mathrm{pr}_2 \circ \sigma) = \gamma. \qquad \square$$

Lemma 4.1.2 *For any two B-modules P_1 and P_2 the direct sum $P := P_1 \oplus P_2$ is relatively projective if and only if P_1 and P_2 both are relatively projective.*

Proof Let pr_1 and pr_2 denote, quite generally, the projection map from a direct sum onto its first and second summand, respectively. Similarly, let i_1, resp. i_2, denote the inclusion map from the first, resp. second, summand into their direct sum.

We first suppose that P is relatively projective. By symmetry it suffices to show that P_1 is relatively projective. We are going to use Lemma 4.1.1. Let therefore $\beta_1 : M_1 \longrightarrow P_1$ be any B-module homomorphism and $\sigma_0 : P_1 \longrightarrow M_1$ be any A-module homomorphism such that $\beta_1 \circ \sigma_0 = \mathrm{id}_{P_1}$. Then

$$\beta\colon \quad M := M_1 \oplus P_2 \xrightarrow{\beta_1 \oplus \mathrm{id}_{P_2}} P_1 \oplus P_2 = P$$

is a B-module homomorphism and

$$\sigma_0'\colon \quad P = P_1 \oplus P_2 \xrightarrow{\sigma_0 \oplus \mathrm{id}_{P_2}} M_1 \oplus P_2 = M$$

is an A-module homomorphism and the two satisfy

$$\beta \circ \sigma_0' = (\beta_1 \circ \sigma_0) \oplus \mathrm{id}_{P_2} = \mathrm{id}_{P_1} \oplus \mathrm{id}_{P_2} = \mathrm{id}_P .$$

By assumption we find a B-module homomorphism $\sigma' : P \longrightarrow M$ such that $\beta \circ \sigma' = \mathrm{id}_P$. We now define the B-module homomorphism $\sigma : P_1 \longrightarrow M_1$ as the composite

$$P_1 \xrightarrow{i_1} P_1 \oplus P_2 \xrightarrow{\sigma'} M_1 \oplus P_2 \xrightarrow{\mathrm{pr}_1} M_1,$$

and we compute

$$\begin{aligned}
\beta_1 \circ \sigma &= \beta_1 \circ \mathrm{pr}_1 \circ \sigma' \circ i_1 = \mathrm{pr}_1 \circ (\beta_1 \oplus \mathrm{id}_{P_2}) \circ \sigma' \circ i_1 \\
&= \mathrm{pr}_1 \circ \beta \circ \sigma' \circ i_1 = \mathrm{pr}_1 \circ i_1 \\
&= \mathrm{id}_{P_1} .
\end{aligned}$$

We now assume, vice versa, that P_1 and P_2 are relatively projective. Again using Lemma 4.1.1 we let $\beta : M \longrightarrow P$ and $\sigma_0 : P \longrightarrow M$ be a B-module and an A-module homomorphism, respectively, such that $\beta \circ \sigma_0 = \mathrm{id}_P$. This latter relation implies that $\sigma_{0,j} := \sigma_0 | P_j$ can be viewed as an A-module homomorphism

$$\sigma_{0,j}\colon \quad P_j \longrightarrow \beta^{-1}(P_j) \quad \text{for } j = 1, 2.$$

We also define the B-module homomorphism $\beta_j := \beta | \beta^{-1}(P_j) : \beta^{-1}(P_j) \longrightarrow P_j$. We obviously have

$$\beta_j \circ \sigma_{0,j} = \mathrm{id}_{P_j} \quad \text{for } j = 1, 2.$$

By assumption we therefore find B-module homomorphisms $\sigma_j : P_j \longrightarrow \beta^{-1}(P_j)$ such that $\beta_j \circ \sigma_j = \mathrm{id}_{P_j}$ for $j = 1, 2$. The B-module homomorphism

$$\begin{aligned}\sigma\colon\quad P = P_1 \oplus P_2 &\longrightarrow M \\ y_1 + y_2 &\longmapsto \sigma_1(y_1) + \sigma_2(y_2)\end{aligned}$$

then satisfies

$$\begin{aligned}\beta \circ \sigma(y_1 + y_2) &= \beta\big(\sigma_1(y_1) + \sigma_2(y_2)\big) = \beta \circ \sigma_1(y_1) + \beta \circ \sigma_2(y_2) \\ &= \beta_1 \circ \sigma_1(y_1) + \beta_2 \circ \sigma_2(y_2) \\ &= y_1 + y_2\end{aligned}$$

for any $y_j \in P_j$. □

Lemma 4.1.3 *For any A-module L_0 the B-module $B \otimes_A L_0$ is relatively projective.*

Proof Consider any pair

$$\begin{array}{ccc} & & B \otimes_A L_0 \\ & & \big\downarrow{\scriptstyle\gamma} \\ M & \xrightarrow{\ \beta\ } & N \end{array}$$

of B-module homomorphisms together with an A-module homomorphism α_0 : $B \otimes_A L_0 \longrightarrow M$ such that $\beta \otimes \alpha_0 = \gamma$. Then

$$\begin{aligned}\alpha\colon\quad B \otimes_A L_0 &\longrightarrow M \\ b \otimes y_0 &\longmapsto b\alpha_0(1 \otimes y_0)\end{aligned}$$

is a B-module homomorphism satisfying $\beta \circ \alpha = \gamma$ since

$$\begin{aligned}\beta \circ \alpha(b \otimes y_0) &= \beta\big(b\alpha_0(1 \otimes y_0)\big) \\ &= b(\beta \circ \alpha_0)(1 \otimes y_0) = b\gamma(1 \otimes y_0) \\ &= \gamma(b \otimes y_0).\end{aligned}$$

□

Proposition 4.1.4 *For any B-module P the following conditions are equivalent:*

i. *P is relatively projective;*
ii. *P is isomorphic to a direct summand of $B \otimes_A P$;*
iii. *P is isomorphic to a direct summand of $B \otimes_A L_0$ for some A-module L_0.*

Proof For the implication from i. to ii. we observe that we have the B-module homomorphism

$$\begin{aligned}\beta\colon\quad B \otimes_A P &\longrightarrow P \\ b \otimes y &\longmapsto by\end{aligned}$$

as well as the A-module homomorphism

$$\sigma_0 \colon \quad P \longrightarrow B \otimes_A P$$
$$y \longmapsto 1 \otimes y$$

which satisfy $\beta \circ \sigma_0 = \mathrm{id}_P$. Hence by Lemma 4.1.1 there exists a B-module homomorphism $\sigma : P \longrightarrow B \otimes_A P$ such that $\beta \circ \sigma = \mathrm{id}_P$. Then $B \otimes_A P = \mathrm{im}(\sigma) \oplus \ker(\beta)$, and $P \xrightarrow{\sigma} \mathrm{im}(\sigma)$ is an isomorphism of B-modules.

The implication from ii. to iii. is trivial. That iii. implies i. follows from Lemmas 4.1.2 and 4.1.3. □

We are primarily interested in the case of an inclusion homomorphism $R[H] \subseteq R[G]$, where R is any commutative ring and H is a subgroup of a finite group G. An $R[G]$-module which is relatively projective with respective to this inclusion will be called *relative $R[H]$-projective.*

Example If $R = k$ is a field then a $k[G]$-module is relatively $k[\{1\}]$-projective if and only if it is projective.

We recall the useful fact that $R[G]$ is free as an $R[H]$-module with basis any set of representatives of the cosets of H in G.

Lemma 4.1.5 *An $R[G]$-module is projective if and only if it is relatively $R[H]$-projective and it is projective as an $R[H]$-module.*

Proof By our initial Remark it remains to show that any projective $R[G]$-module P also is projective as an $R[H]$-module. But P is a direct summand of a free $R[G]$-module $F \cong \bigoplus_{i \in I} R[G]$ by Proposition 1.6.4. Since $R[G]$ is free as an $R[H]$-module F also is free as an $R[H]$-module. Hence the reverse implication in Proposition 1.6.4 says that P is projective as an $R[H]$-module. □

As we have discussed in Sect. 2.3 the $R[G]$-module $R[G] \otimes_{R[H]} L_0$, for any $R[H]$-module L_0, is naturally isomorphic to the induced module $\mathrm{Ind}_H^G(L_0)$. Hence we may restate Proposition 4.1.4 as follows.

Proposition 4.1.6 *For any $R[G]$-module P the following conditions are equivalent*:

i. *P is relatively $R[H]$-projective*;
ii. *P is isomorphic to a direct summand of* $\mathrm{Ind}_H^G(P)$;
iii. *P is isomorphic to a direct summand of* $\mathrm{Ind}_H^G(L_0)$ *for some $R[H]$-module L_0.*

Next we give a useful technical criterion.

Lemma 4.1.7 *Let $\{g_1, \ldots, g_m\} \subseteq G$ be a set of representatives for the left cosets of H in G; a left $R[G]$-module P is relatively $R[H]$-projective if and only if there exists an $R[H]$-module endomorphism $\psi : P \longrightarrow P$ such that*

$$\sum_{i=1}^{m} g_i \psi g_i^{-1} = \mathrm{id}_P .$$

Proof We introduce the following notation. For any $g \in G$ there are uniquely determined elements $h_1(g), \ldots, h_m(g) \in H$ and a uniquely determined permutation $\pi(g, .)$ of $\{1, \ldots, m\}$ such that

$$g g_i = g_{\pi(g,i)} h_i(g) \quad \text{for any } 1 \le i \le m.$$

Step 1: Let $\alpha_0 : M \longrightarrow N$ be any $R[H]$-module homomorphism between any two $R[G]$-modules. We claim that

$$\begin{aligned} \alpha := \mathcal{N}(\alpha_0): \quad M &\longrightarrow N \\ x &\longmapsto \sum_{i=1}^{m} g_i \alpha_0\big(g_i^{-1} x\big) \end{aligned}$$

is an $R[G]$-module homomorphism. Let $g \in G$. We compute

$$\begin{aligned} \alpha\big(g^{-1}x\big) &= \sum_{i=1}^{m} g_i \alpha_0\big(g_i^{-1} g^{-1} x\big) = \sum_{i=1}^{m} g_i \alpha_0\big((g g_i)^{-1} x\big) \\ &= \sum_{i=1}^{m} g_i \alpha_0\big(\big(g_{\pi(g,i)} h_i(g)\big)^{-1} x\big) = \sum_{i=1}^{m} g_i \alpha_0\big(h_i(g)^{-1} g_{\pi(g,i)}^{-1} x\big) \\ &= \sum_{i=1}^{m} g_i h_i(g)^{-1} \alpha_0\big(g_{\pi(g,i)}^{-1} x\big) = \sum_{i=1}^{m} g^{-1} g_{\pi(g,i)} \alpha_0\big(g_{\pi(g,i)}^{-1} x\big) \\ &= g^{-1} \alpha(x) \end{aligned}$$

for any $x \in M$ which proves the claim. We also observe that, for any $R[G]$-module homomorphisms $\beta : M' \longrightarrow M$ and $\gamma : N \longrightarrow N'$, we have

$$\gamma \circ \mathcal{N}(\alpha_0) \circ \beta = \mathcal{N}(\gamma \circ \alpha_0 \circ \beta). \tag{4.1.1}$$

Step 2: Let us temporarily say that an $R[H]$-module endomorphism $\psi : M \longrightarrow M$ of an $R[G]$-module M is *good* if $\mathcal{N}(\psi) = \mathrm{id}_M$ holds true. The assertion of the present lemma then says that an $R[G]$-module is relatively $R[H]$-projective if and only if it has a good endomorphism. In this step we establish that any

$R[G]$-module P with a good endomorphism ψ is relatively $R[H]$-projective. Let

$$\begin{array}{ccc} & & P \\ & & \downarrow \gamma \\ M & \xrightarrow{\beta} & N \end{array}$$

be a pair of $R[G]$-module homomorphisms and $\alpha_0 : P \longrightarrow M$ be an $R[H]$-module homomorphism such that $\beta \circ \alpha_0 = \gamma$. Using (4.1.1) we see that $\alpha := \mathcal{N}(\alpha_0 \circ \psi)$ satisfies

$$\beta \circ \alpha = \beta \circ \mathcal{N}(\alpha_0 \circ \psi) = \mathcal{N}(\beta \circ \alpha_0 \circ \psi) = \mathcal{N}(\gamma \circ \psi) = \gamma \circ \mathcal{N}(\psi) = \mathrm{id}_P.$$

Step 3: If the direct sum $M = M_1 \oplus M_2$ of $R[G]$-modules has a good endomorphism ψ then each summand M_j has a good endomorphism as well. Let $M_j \xrightarrow{i_j} M \xrightarrow{\mathrm{pr}_j} M_j$ be the inclusion and the projection map, respectively. Then $\psi_j := \mathrm{pr}_j \circ \psi \circ i_j$ satisfies

$$\mathcal{N}(\psi_j) = \mathcal{N}(\mathrm{pr}_j \circ \psi \circ i_j) = \mathrm{pr}_j \circ \mathcal{N}(\psi) \circ i_j = \mathrm{pr}_j \circ i_j = \mathrm{id}_{M_j}$$

where, for the second equality, we again have used (4.1.1).

Step 4: In order to show that any relatively $R[H]$-projective $R[G]$-module P has a good endomorphism it remains, by Step 3 and Proposition 4.1.4, to see that any $R[G]$-module of the form $R[G] \otimes_{R[H]} L_0$, for some $R[H]$-module L_0, has a good endomorphism. We have

$$R[G] \otimes_{R[H]} L_0 = g_1 \otimes L_0 + \cdots + g_m \otimes L_0$$

and, assuming that $g_1 \in H$, we define the map

$$\begin{aligned} \psi : \quad R[G] \otimes_{R[H]} L_0 &\longrightarrow R[G] \otimes_{R[H]} L_0 \\ \sum_{i=1}^{m} g_i \otimes x_i &\longmapsto g_1 \otimes x_1. \end{aligned}$$

Let $h \in H$ and observe that the permutation $\pi(h, .)$ has the property that $\pi(h, 1) = 1$. We compute

$$\begin{aligned} \psi\left(h\left(\sum_{i=1}^{m} g_i \otimes x_i\right)\right) &= \psi\left(\sum_{i=1}^{m} hg_i \otimes x_i\right) = \psi\left(\sum_{i=1}^{m} g_{\pi(h,i)} h_i(h) \otimes x_i\right) \\ &= \psi\left(\sum_{i=1}^{m} g_{\pi(h,i)} \otimes h_i(h) x_i\right) = g_{\pi(h,1)} \otimes h_1 x_1 \end{aligned}$$

$$= g_{\pi(h,1)} h_1(h) \otimes x_1 = h g_1 \otimes x_1$$
$$= h\psi\left(\sum_{i=1}^{m} g_i \otimes x_i\right).$$

This shows that ψ is an $R[H]$-module endomorphism. Moreover, using that $\pi(g_j^{-1}, i) = 1$ if and only if $i = j$, we obtain

$$\begin{aligned}
&\sum_{j=1}^{m} g_j \psi\left(g_j^{-1} \sum_{i=1}^{m} g_i \otimes x_i\right) \\
&\quad = \sum_{j=1}^{m} g_j \psi\left(\sum_{i=1}^{m} g_j^{-1} g_i \otimes x_i\right) \\
&\quad = \sum_{j=1}^{m} g_j \psi\left(\sum_{i=1}^{m} g_{\pi(g_j^{-1}, i)} h_i(g_j^{-1}) \otimes x_i\right) \\
&\quad = \sum_{j=1}^{m} g_j \psi\left(\sum_{i=1}^{m} g_{\pi(g_j^{-1}, i)} \otimes h_i(g_j^{-1}) x_i\right) \\
&\quad = \sum_{j=1}^{m} g_j g_{\pi(g_j^{-1}, j)} \otimes h_j(g_j^{-1}) x_j = \sum_{j=1}^{m} g_j g_{\pi(g_j^{-1}, j)} h_j(g_j^{-1}) \otimes x_j \\
&\quad = \sum_{j=1}^{m} g_j g_j^{-1} g_j \otimes x_j = \sum_{j=1}^{m} g_j \otimes x_j.
\end{aligned}$$

Hence ψ is good. □

If k is a field of characteristic $p > 0$ and the order of the group G is prime to p then we know the group ring $k[G]$ to be semisimple. Hence all $k[G]$-modules are projective (cf. Remark 1.7.3). This fact generalizes as follows to the relative situation.

Proposition 4.1.8 *If the integer $[G : H]$ is invertible in R then any $R[G]$-module M is relatively $R[H]$-projective.*

Proof The endomorphism $\frac{1}{[G:H]} \mathrm{id}_M$ of M is good. □

4.2 Vertices and Sources

For any $R[G]$-module M we introduce the set

$$\mathcal{V}(M) := \text{set of subgroups } H \subseteq G \quad \text{such that} \\ M \text{ is relatively } \mathbb{R}[H]\text{-projective.}$$

For trivial reasons G lies in $\mathcal{V}(M)$.

Lemma 4.2.1 *The set $\mathcal{V}(M)$ is closed under conjugation, i.e. for any $H \in \mathcal{V}(M)$ and $g \in G$ we have $gHg^{-1} \in \mathcal{V}(M)$.*

Proof Let

$$\begin{array}{ccc} & & M \\ & & \downarrow \gamma \\ L & \xrightarrow{\beta} & N \end{array}$$

be any pair of $R[G]$-module homomorphisms and $\alpha_0 : M \longrightarrow L$ be an $R[gHg^{-1}]$-module homomorphism such that $\beta \circ \alpha_0 = \gamma$. We consider the map $\alpha_1 := g^{-1}\alpha_0 g : M \longrightarrow L$. For $h \in H$ we have

$$\alpha_1(hy) = g^{-1}\alpha_0(ghy) = g^{-1}\alpha_0\big(ghg^{-1}gy\big) = g^{-1}ghg^{-1}\alpha_0(gy) = h\alpha_1(y)$$

for any $y \in M$. This shows that α_1 is an $R[H]$-module homomorphism. It satisfies

$$\begin{aligned} \beta \circ \alpha_1(y) &= \beta\big(g^{-1}\alpha_0(gy)\big) = g^{-1}\big(\beta \circ \alpha_0(gy)\big) \\ &= g^{-1}\gamma(gy) = \gamma\big(g^{-1}gy\big) \\ &= \gamma(y) \end{aligned}$$

for any $y \in M$. Since, by assumption, M is relatively $R[H]$-projective there exists an $R[G]$-module homomorphism $\alpha : M \longrightarrow L$ such that $\beta \circ \alpha = \gamma$. □

We let $\mathcal{V}_0(M) \subseteq \mathcal{V}(M)$ denote the subset of those subgroups $H \in \mathcal{V}(M)$ which are minimal with respect to inclusion.

Exercise 4.2.2

i. $\mathcal{V}_0(M)$ is nonempty.
ii. $\mathcal{V}_0(M)$ is closed under conjugation.
iii. $\mathcal{V}(M)$ is the set of all subgroups of G which contain some subgroup in $\mathcal{V}_0(M)$.

Definition A subgroup $H \in \mathcal{V}_0(M)$ is called a vertex of M.

Lemma 4.2.3 *Let p be a fixed prime number and suppose that all prime numbers $\neq p$ are invertible in R (e.g., let $R = k$ be a field of characteristic p or let R be a $(0, p)$-ring for such a field k); then all vertices are p-groups.*

Proof Let $H_1 \in \mathcal{V}_0(M)$ for some $R[G]$-module M and let $H_0 \subseteq H_1$ be a p-Sylow subgroup of H_1. We claim that M is relatively $R[H_0]$-projective which, by the minimality of H_1, implies that $H_0 = H_1$ is a p-group. Let

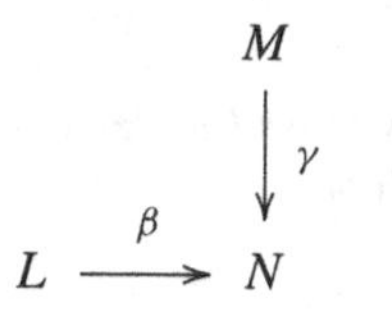

be a pair of $R[G]$-module homomorphisms together with an $R[H_0]$-module homomorphism $\alpha_0 : M \longrightarrow L$ such that $\beta \circ \alpha_0 = \gamma$. Since $[H_1 : H_0]$ is invertible in R it follows from Proposition 4.1.8 that M viewed as an $R[H_1]$-module is relatively $R[H_0]$-projective. Hence there exists an $R[H_1]$-module homomorphism $\alpha_1 : M \longrightarrow L$ such that $\beta \circ \alpha_1 = \gamma$. But $H_1 \in \mathcal{V}(M)$. So there further must exist an $R[G]$-module homomorphism $\alpha : M \longrightarrow L$ satisfying $\beta \circ \alpha = \gamma$. □

In order to investigate vertices more closely we need a general property of induction. To formulate it we first generalize some notation from Sect. 2.5. Let $H \subseteq G$ be a subgroup and $g \in G$ be an element. For any $R[gHg^{-1}]$-module M, corresponding to the ring homomorphism $\pi : R[gHg^{-1}] \longrightarrow \mathrm{End}_R(M)$, we introduce the $R[H]$-module g^*M defined by the composite ring homomorphism

$$\begin{aligned} R[H] &\longrightarrow R\big[gHg^{-1}\big] \xrightarrow{\pi} \mathrm{End}_R(M) \\ h &\longmapsto ghg^{-1}. \end{aligned}$$

Proposition 4.2.4 (Mackey) *Let $H_0, H_1 \subseteq G$ be two subgroups and fix a set $\{g_1, \dots, g_m\} \subseteq G$ of representatives for the double cosets in $H_0 \backslash G / H_1$. For any $R[H_1]$-module L we have an isomorphism of $R[H_0]$-modules*

$$\mathrm{Ind}_{H_1}^G(L) \cong \mathrm{Ind}_{H_0 \cap g_1 H_1 g_1^{-1}}^{H_0}\big(\big(g_1^{-1}\big)^* L\big) \oplus \cdots \oplus \mathrm{Ind}_{H_0 \cap g_m H_1 g_m^{-1}}^{H_0}\big(\big(g_m^{-1}\big)^* L\big).$$

Proof As an H_0-set through left multiplication and simultaneously as a (right) H_1-set through right multiplication G decomposes disjointly into

$$G = H_0 g_1 H_1 \cup \cdots \cup H_0 g_m H_1.$$

Therefore the induced module $\mathrm{Ind}_{H_1}^G(L)$, viewed as an $R[H_0]$-module, decomposes into the direct sum of $R[H_0]$-modules

$$\mathrm{Ind}_{H_1}^G(L) = \mathrm{Ind}_{H_1}^{H_0 g_1 H_1}(L) \oplus \cdots \oplus \mathrm{Ind}_{H_1}^{H_0 g_m H_1}(L)$$

where, for any $g \in G$, we put

$$\mathrm{Ind}_{H_1}^{H_0 g H_1}(L) := \{\phi \in \mathrm{Ind}_{H_1}^{G}(L) : \phi(g') = 0 \text{ for any } g' \notin H_0 g H_1\}.$$

We note that $\mathrm{Ind}_{H_1}^{H_0 g H_1}(L)$ is the $R[H_0]$-module of all functions $\phi : H_0 g H_1 \to L$ such that $\phi(h_0 g h_1) = h_1^{-1}\phi(h_0 g)$ for any $h_0 \in H_0$ and $h_1 \in H_1$. It remains to check that the map

$$\mathrm{Ind}_{H_0 \cap g H_1 g^{-1}}^{H_0}\big((g^{-1})^* L\big) \longrightarrow \mathrm{Ind}_{H_1}^{H_0 g H_1}(L)$$

$$\phi \longmapsto \phi^\sharp(h_0 g h_1) := h_1^{-1}\phi(h_0),$$

where $h_0 \in H_0$ and $h_1 \in H_1$, is an isomorphism of $R[H_0]$-modules. In order to check that $\phi^\sharp$ is well defined, let $h_0 g h_1 = \tilde{h}_0 g \tilde{h}_1$ with $\tilde{h}_i \in H_i$. Then

$$h_0 = \tilde{h}_0\big(g\tilde{h}_1 h_1^{-1} g^{-1}\big) \quad \text{with } g\tilde{h}_1 h_1^{-1} g^{-1} \in H_0 \cap g H_1 g^{-1}$$

and hence

$$\phi(h_0) = \phi\big(\tilde{h}_0\big(g\tilde{h}_1 h_1^{-1} g^{-1}\big)\big) = \big(\tilde{h}_1 h_1^{-1}\big)^{-1}\phi(\tilde{h}_0)$$

which implies

$$\phi^\sharp(h_0 g h_1) = h_1^{-1}\phi(h_0) = \tilde{h}_1^{-1}\phi(\tilde{h}_0) = \phi^\sharp(\tilde{h}_0 g \tilde{h}_1).$$

It is clear that the map $\phi^\sharp : H_0 g H_1 \longrightarrow L$ lies in $\mathrm{Ind}_{H_1}^{H_0 g H_1}(L)$. For $h \in H_0$ we compute

$$\begin{aligned}({}^h\phi)^\sharp(h_0 g h_1) &= h_1^{-1}{}^h\phi(h_0) = h_1^{-1}\phi(h^{-1}h_0)\\ &= \phi^\sharp(h^{-1}h_0 g h_1)\\ &= {}^h(\phi^\sharp)(h_0 g h_1).\end{aligned}$$

This shows that $\phi \longmapsto \phi^\sharp$ is an $R[H_0]$-module homomorphism. It is visibly injective. To establish its surjectivity we let $\psi \in \mathrm{Ind}_{H_1}^{H_0 g H_1}(L)$ and we define the map

$$\begin{aligned}\phi\colon\ H_0 &\longrightarrow L\\ h_0 &\longmapsto \psi(h_0 g).\end{aligned}$$

For $h \in H_0 \cap g H_1 g^{-1}$ we compute

$$\begin{aligned}\phi(h_0 h) &= \psi(h_0 h g) = \psi(h_0 g g^{-1} h g) = (g^{-1} h g)^{-1}\psi(h_0 g)\\ &= g^{-1}h^{-1}g\phi(h_0)\end{aligned}$$

which means that $\phi \in \mathrm{Ind}_{H_0 \cap gH_1g^{-1}}^{H_0}((g^{-1})^*L)$. We obviously have

$$\phi^\sharp(h_0gh_1) = h_1^{-1}\phi(h_0) = h_1^{-1}\psi(h_0g) = \psi(h_0gh_1),$$

i.e. $\phi^\sharp = \psi$. □

Proposition 4.2.5 *Suppose that R is noetherian and complete and that $R/\mathrm{Jac}(R)$ is artinian. For any finitely generated and indecomposable $R[G]$-module M its set of vertices $\mathcal{V}_0(M)$ consists of a single conjugacy class of subgroups.*

Proof We know from Exercise 4.2.2.ii that $\mathcal{V}_0(M)$ is a union of conjugacy classes. In the following we show that any two $H_0, H_1 \in \mathcal{V}_0(M)$ are conjugate. By Proposition 4.1.6 the $R[G]$-module M is isomorphic to a direct summand of $\mathrm{Ind}_{H_0}^G(M)$ as well as of $\mathrm{Ind}_{H_1}^G(M)$. The latter implies that $\mathrm{Ind}_{H_0}^G(M)$ is isomorphic to a direct summand of $\mathrm{Ind}_{H_0}^G(\mathrm{Ind}_{H_1}^G(M))$. Together with the former we obtain that M is isomorphic to a direct summand of $\mathrm{Ind}_{H_0}^G(\mathrm{Ind}_{H_1}^G(M))$. At this point we induce the $R[H_0]$-module decomposition

$$\mathrm{Ind}_{H_1}^G(M) \cong \bigoplus_{i=1}^m \mathrm{Ind}_{H_0 \cap g_iH_1g_i^{-1}}^{H_0}\big((g_i^{-1})^*M\big)$$

from Mackey's Proposition 4.2.4 to G and get, by transitivity of induction, that

$$\mathrm{Ind}_{H_0}^G\big(\mathrm{Ind}_{H_1}^G(M)\big) \cong \bigoplus_{i=1}^m \mathrm{Ind}_{H_0 \cap g_iH_1g_1^{-1}}^{G}\big((g_i^{-1})^*M\big)$$

as $R[G]$-modules. Our assumptions on R guarantee the validity of the Krull–Remak–Schmidt Theorem 1.4.7 for $R[G]$, i.e. the "unicity" of the decomposition into indecomposable modules of the finitely generated $R[G]$-modules $\mathrm{Ind}_{H_0}^G(\mathrm{Ind}_{H_1}^G(M))$ and $\mathrm{Ind}_{H_0 \cap g_iH_1g_i^{-1}}^G((g_i^{-1})^*M)$. It follows that any indecomposable direct summand of $\mathrm{Ind}_{H_0}^G(\mathrm{Ind}_{H_0}^G(M))$, for example our M, must be a direct summand of at least one of the $\mathrm{Ind}_{H_0 \cap g_iH_1g_i^{-1}}^G((g_i^{-1})^*M)$. This implies, by Proposition 4.1.6, that M is relatively $R[H_0 \cap gH_1g^{-1}]$-projective for some $g \in G$. Hence $H_0 \cap gH_1g^{-1} \in \mathcal{V}(M)$. Since $\mathcal{V}(M)$ is closed under conjugation according to Lemma 4.2.1 we also have $g^{-1}H_0g \cap H_1 \in \mathcal{V}(M)$. Since H_0 and H_1 are both minimal in $\mathcal{V}(M)$ we must have $H_0 = gH_1g^{-1}$. □

Definition Suppose that M is finitely generated and indecomposable, and let $V \in \mathcal{V}_0(M)$. Any finitely generated indecomposable $R[V]$-module Q such that M is isomorphic to a direct summand of $\mathrm{Ind}_V^G(Q)$ is called a V-source of M. A source of M is a V-source for some $V \in \mathcal{V}_0(M)$.

We remind the reader that $N_G(H) = \{g \in G : gHg^{-1} = H\}$ denotes the normalizer of the subgroup H of G. Since G is finite any of the inclusions $gHg^{-1} \subseteq H$ or $gHg^{-1} \supseteq H$ already implies $g \in N_G(H)$.

Proposition 4.2.6 *Suppose that R is noetherian and complete and that $R/\operatorname{Jac}(R)$ is artinian. Let M be a finitely generated indecomposable $R[G]$-module and let $V \in \mathcal{V}_0(M)$. We then have:*

i. *M has a V-source Q which is a direct summand of M as an $R[V]$-module;*
ii. *if Q is a V-source of M then $(g^{-1})^* Q$, for any $g \in G$, is a gVg^{-1}-source of M;*
iii. *for any two V-sources Q_0 and Q_1 of M there exists a $g \in N_G(V)$ such that $Q_1 \cong g^* Q_0$.*

Proof i. We decompose M, as an $R[V]$-module, into a direct sum

$$M = M_1 \oplus \cdots \oplus M_r$$

of indecomposable $R[V]$-modules M_i. Then

$$\operatorname{Ind}_V^G(M) = \bigoplus_{i=1}^{r} \operatorname{Ind}_V^G(M_i).$$

By Proposition 4.1.6 the indecomposable $R[G]$-module M is isomorphic to a direct summand of $\operatorname{Ind}_V^G(M)$. As noted already in the proof of Proposition 4.2.5 the Krull–Remak–Schmidt Theorem 1.4.7 implies that M then has to be isomorphic to a direct summand of some $\operatorname{Ind}_V^G(M_{i_0})$. Put $Q := M_{i_0}$.

ii. That Q is a finitely generated indecomposable $R[V]$-module implies that $(g^{-1})^* Q$ is a finitely generated indecomposable $R[gVg^{-1}]$-module. The map

$$\begin{aligned} \operatorname{Ind}_V^G(Q) &\xrightarrow{\cong} \operatorname{Ind}_{gVg^{-1}}^G\big((g^{-1})^* Q\big) \\ \phi &\longmapsto \phi(.g) \end{aligned}$$

is an isomorphism of $R[G]$-modules. Hence if M is isomorphic to a direct summand of the former it also is isomorphic to a direct summand of the latter.

iii. By i. we may assume that Q_0 is a direct summand of M as an $R[V]$-module. But, by assumption, M, as an $R[G]$-module, and therefore also Q_0, as an $R[V]$-module, is isomorphic to a direct summand of $\operatorname{Ind}_V^G(Q_1)$ as well as of $\operatorname{Ind}_V^G(Q_0)$. This implies that Q_0 is isomorphic to a direct summand of $\operatorname{Ind}_V^G(\operatorname{Ind}_V^G(Q_1))$. By the same reasoning with Mackey's Proposition 4.2.4 as in the proof of Proposition 4.2.5 we deduce that

$$Q_0 \text{ is isomorphic to a direct summand of } \operatorname{Ind}_{V \cap gVg^{-1}}^V\big((g^{-1})^* Q_1\big) \tag{4.2.1}$$

for some $g \in G$. Then $\operatorname{Ind}_V^G(Q_0)$ and hence also M are isomorphic to a direct summand of $\operatorname{Ind}_{V \cap gVg^{-1}}^G((g^{-1})^* Q_1)$. Since V is a vertex of M we must have

$|V| \le |V \cap gVg^{-1}|$, hence $V \subseteq gVg^{-1}$, and therefore $g \in N_G(V)$. By inserting this information into (4.2.1) it follows that Q_0 is isomorphic to a direct summand of $(g^{-1})^* Q_1$. But with Q_1 also $(g^{-1})^* Q_1$ is indecomposable. We finally obtain $Q_0 \cong (g^{-1})^* Q_1$. □

Exercise 4.2.7 If Q is a V-source of M then $\mathcal{V}(Q) = \mathcal{V}_0(Q) = \{V\}$.

4.3 The Green Correspondence

Throughout this section we assume the commutative ring R to be noetherian and complete and $R/\operatorname{Jac}(R)$ to be artinian. We fix a subgroup $H \subseteq G$.

We consider any finitely generated indecomposable $R[G]$-module M such that $H \in \mathcal{V}(M)$. Let

$$M = L_1 \oplus \cdots \oplus L_r$$

be a decomposition of M as an $R[H]$-module into indecomposable $R[H]$-modules L_i. The Krull–Remak–Schmidt Theorem 1.4.7 says that the set

$$\mathrm{IND}_H\big(\{M\}\big) := \big\{\{L_1\}, \ldots, \{L_r\}\big\}$$

of isomorphism classes of the $R[H]$-modules L_i only depends on the isomorphism class $\{M\}$ of the $R[G]$-module M.

Lemma 4.3.1 *Let $V \in \mathcal{V}_0(M)$ and $V_i \in \mathcal{V}_0(L_i)$ for $1 \le i \le r$; we then have:*

i. *For any $1 \le i \le r$ there is a $g_i \in G$ such that $V_i \subseteq g_i V g_i^{-1}$;*
ii. *M is isomorphic to a direct summand of $\operatorname{Ind}_H^G(L_i)$ for some $1 \le i \le r$;*
iii. *if M is isomorphic to a direct summand of $\operatorname{Ind}_H^G(L_i)$ then $V_i = g_i V g_i^{-1}$ and, in particular, $\mathcal{V}_0(L_i) \subseteq \mathcal{V}_0(M)$;*
iv. *if $\mathcal{V}_0(L_i) \subseteq \mathcal{V}_0(M)$ then M and L_i have a common V_i-source.*

(The assertions i. *and* iv. *do not require the assumption that $H \in \mathcal{V}(M)$.)*

Proof We fix a V-source Q of M. Then M, as an $R[G]$-module, is isomorphic to a direct summand of $\operatorname{Ind}_V^G(Q)$. Using Mackey's Proposition 4.2.4 we see that $L_1 \oplus \cdots \oplus L_r$ is isomorphic to a direct summand of

$$\bigoplus_{x \in \mathcal{R}} \operatorname{Ind}_{H \cap xVx^{-1}}^H\big((x^{-1})^* Q\big)$$

where $\mathcal{R} \subseteq G$ is a fixed set of representatives for the double cosets in $H \backslash G / V$. We conclude from the Krull–Remak–Schmidt Theorem 1.4.7 that, for any $1 \le i \le r$, there exists an $x_i \in \mathcal{R}$ such that L_i is isomorphic to a direct summand of $\operatorname{Ind}_{H \cap x_i V x_i^{-1}}^H((x_i^{-1})^* Q)$. Hence $H \cap x_i V x_i^{-1} \in \mathcal{V}(L_i)$ by Proposition 4.1.6 and, since $\mathcal{V}_0(L_i)$ is a single conjugacy class with respect to H by Proposition 4.2.5, we have $hV_ih^{-1} \subseteq H \cap x_i V x_i^{-1}$ for some $h \in H$. We put $g_i := h^{-1}x_i$ and obtain

$$V_i \subseteq g_i V g_i^{-1}$$

which proves i.

Suppose that $V_i \in \mathcal{V}_0(M)$. Then $|V_i| = |V|$ (cf. Proposition 4.2.5) and hence

$$V_i = g_i V g_i^{-1} = h^{-1} x_i V x_i^{-1} h \subseteq H.$$

In particular, $x_i V x_i^{-1}$ is a vertex of L_i and $(x_i^{-1})^* Q$ is an $x_i V x_i^{-1}$-source of L_i. Using Proposition 4.2.6.ii we deduce that $(g_i^{-1})^* Q$ is a V_i-source of L_i. On the other hand we have, of course, that $g_i V g_i^{-1} \in \mathcal{V}_0(M)$, and Proposition 4.2.6.ii again implies that $(g_i^{-1})^* Q$ is a V_i-source of M. This shows iv.

Since $H \in \mathcal{V}(M)$ the $R[G]$-module M, by Proposition 4.1.6, also is isomorphic to a direct summand of

$$\operatorname{Ind}_H^G(M) = \bigoplus_{i=1}^r \operatorname{Ind}_H^G(L_i).$$

Hence ii. follows from the Krull–Remak–Schmidt Theorem 1.4.7. Now suppose that M, as an $R[G]$-module, is isomorphic to a direct summand of $\operatorname{Ind}_H^G(L_i)$ for some $1 \le i \le r$. Let Q_i be a V_i-source of L_i. Then $\operatorname{Ind}_H^G(L_i)$, and hence M, is isomorphic to a direct summand of $\operatorname{Ind}_{V_i}^G(Q_i)$. It follows that $V_i \in \mathcal{V}(M)$ and therefore that $|V| \le |V_i|$. Together with i. we obtain $V_i = g_i V g_i^{-1}$. This proves iii. □

We introduce the set

$$\mathcal{V}_0^H(M) := \{V \in \mathcal{V}_0(M) : V \subseteq H\}.$$

Obviously the subgroup H acts on $\mathcal{V}_0^H(M)$ by conjugation so that we can speak of the H-orbits in $\mathcal{V}_0^H(M)$. We also introduce

$$\mathrm{IND}_H^0(\{M\}) := \{\{L_i\} \in \mathrm{IND}_H(\{M\}) : \mathcal{V}_0(L_i) \subseteq \mathcal{V}_0(M)\}$$

and

$$\mathrm{IND}_H^1(\{M\}) := \{\{L_i\} \in \mathrm{IND}_H(\{M\}) : M \text{ is isomorphic to a direct summand of } \operatorname{Ind}_H^G(L_i)\}.$$

By Lemma 4.3.1 all three sets are nonempty and

$$\mathrm{IND}_H^1(\{M\}) \subseteq \mathrm{IND}_H^0(\{M\}).$$

Furthermore, using Proposition 4.2.5 we have the obvious map

$$\begin{aligned} v_H^M : \quad \mathrm{IND}_H^0(\{M\}) &\longrightarrow \text{set of } H\text{-orbits in } \mathcal{V}_0^H(M) \\ \{L_i\} &\longmapsto \mathcal{V}_0(L_i). \end{aligned}$$

Lemma 4.3.2 *For any $V \in \mathcal{V}_0^H(M)$ there is a $1 \leq j \leq r$ such that $V \in \mathcal{V}_0(L_j)$; in this case M and L_j have a common V-source.*

Proof By Proposition 4.2.6.i the $R[G]$-module M has a V-source M_0 which is a direct summand of M as an $R[V]$-module. Since $V \subseteq H$ the Krull–Remak–Schmidt Theorem 1.4.7 implies that M_0 is isomorphic to a direct summand of some L_j as $R[V]$-modules. This means that we have a decomposition

$$L_j = X_1 \oplus \cdots \oplus X_t$$

into indecomposable $R[V]$-modules X_i such that $X_1 \cong M_0$. We note that V is a vertex of X_1 by Exercise 4.2.7. Let V_j be a vertex of L_j. We now apply Lemma 4.3.1.i to this decomposition (i.e. with H, L_j, V_j, V instead of G, M, V, V_1) and obtain that $V \subseteq hV_jh^{-1}$ for some $h \in H$. But Lemma 4.3.1.i applied to the original decomposition $M = L_1 \oplus \cdots \oplus L_r$ gives $|V| \geq |V_j|$. Hence $V = hV_jh^{-1}$ which implies $V \in \mathcal{V}_0(L_j)$. The latter further implies $V_j \in \mathcal{V}_0(L_j) \subseteq \mathcal{V}_0(M)$. Therefore M and L_j, by Lemma 4.3.1.iv, have a common V_j-source and hence, by Proposition 4.2.6.ii, also a common V-source. □

Lemma 4.3.2 implies that the map v_H^M is surjective.

Lemma 4.3.3 *For any $V \in \mathcal{V}_0^H(M)$ there exists a finitely generated indecomposable $R[H]$-module N such that $V \in \mathcal{V}_0(N)$ and M is isomorphic to a direct summand of $\mathrm{Ind}_H^G(N)$.*

Proof Let Q be a V-source of M so that M is isomorphic to a direct summand of $\mathrm{Ind}_V^G(Q) = \mathrm{Ind}_H^G(\mathrm{Ind}_V^H(Q))$. By the usual argument there exists an indecomposable direct summand N of $\mathrm{Ind}_V^H(Q)$ such that M is isomorphic to a direct summand of $\mathrm{Ind}_H^G(N)$. According to Proposition 4.1.6 we have $V \in \mathcal{V}(N)$. We choose a vertex $V' \in \mathcal{V}_0(N)$ such that $V' \subseteq V$. Let Q' be a V'-source of N. Then N is isomorphic to a direct summand of $\mathrm{Ind}_{V'}^H(Q')$. Hence M is isomorphic to a direct summand of $\mathrm{Ind}_H^G(\mathrm{Ind}_{V'}^H(Q')) = \mathrm{Ind}_{V'}^G(Q')$, and consequently $V' \in \mathcal{V}(M)$ by Proposition 4.1.6. The minimality of V implies $V' = V$ which shows that V is a vertex of N. □

Lemma 4.3.4 *Let N be a finitely generated indecomposable $R[H]$-module, and let $U \in \mathcal{V}_0(N)$; we then have:*

i. *The induced module $\mathrm{Ind}_H^G(N)$, as an $R[H]$-module, is of the form*

$$\mathrm{Ind}_H^G(N) \cong N \oplus N_1 \oplus \cdots \oplus N_s$$

with indecomposable $R[H]$-modules N_i for which there are elements $g_i \in G \setminus H$ such that $H \cap g_i U g_i^{-1} \in \mathcal{V}(N_i)$;

ii. $\mathrm{Ind}_H^G(N) = M' \oplus M''$ *with* $R[G]$*-submodules* M' *and* M'' *such that:*

- M' *is indecomposable,*
- N *is isomorphic to a direct summand of* M',
- $U \in \mathcal{V}_0(M')$, *and* M' *and* N *have a common* U*-source;*

iii. *if* $N_G(U) \subseteq H$ *then* $\mathcal{V}_0(N_i) \neq \mathcal{V}_0(N)$ *for any* $1 \leq i \leq s$ *and* N *is not isomorphic to a direct summand of* M''.

Proof i. Since $U \in \mathcal{V}_0(N)$ we have

$$\mathrm{Ind}_U^H(N) \cong N \oplus N'$$

for some $R[H]$-module N'. On the other hand, let

$$\mathrm{Ind}_H^G(N') = N_0 \oplus N_1 \oplus \cdots \oplus N_s$$

be a decomposition into indecomposable $R[H]$-modules N_i. Furthermore, it is easy to see (but can also be deduced from Mackey's Proposition 4.2.4) that, as $R[H]$-modules, we have

$$\mathrm{Ind}_H^G(N') \cong N' \oplus N''$$

for some $R[H]$-module N'', and correspondingly that N is isomorphic to a direct summand of $\mathrm{Ind}_H^G(N)$. The latter implies that N is isomorphic to one of the N_i. We may assume that $N \cong N_0$. Combining these facts with Mackey's Proposition 4.2.4 we obtain

$$\begin{aligned}
N \oplus N_1 \oplus \cdots \oplus N_s \oplus N' \oplus N''& \\
&\cong \mathrm{Ind}_H^G(N) \oplus \mathrm{Ind}_H^G(N') \\
&\cong \mathrm{Ind}_H^G(\mathrm{Ind}_U^H(N)) = \mathrm{Ind}_U^G(N) \\
&\cong \bigoplus_{g \in \mathcal{R}} \mathrm{Ind}_{H \cap gUg^{-1}}^H((g^{-1})^* N) \\
&= \mathrm{Ind}_U^H(N) \oplus \Big(\bigoplus_{g \in \mathcal{R}, g \notin H} \mathrm{Ind}_{H \cap gUg^{-1}}^H((g^{-1})^* N) \Big) \\
&\cong N \oplus N' \oplus \Big(\bigoplus_{g \in \mathcal{R}, g \notin H} \mathrm{Ind}_{H \cap gUg^{-1}}^H((g^{-1})^* N) \Big)
\end{aligned}$$

as $R[H]$-modules, where $\mathcal{R}$ is a fixed set of representatives for the double cosets in $H \backslash G / U$ such that $1 \in \mathcal{R}$. It is a consequence of the Krull–Remak–Schmidt Theorem 1.4.7 that we may cancel summands which occur on both sides and still have an isomorphism

$$N_1 \oplus \cdots \oplus N_s \oplus N'' \cong \bigoplus_{g \in \mathcal{R}, g \notin H} \mathrm{Ind}_{H \cap gUg^{-1}}^H((g^{-1})^* N).$$

Moreover, it follows that N_i, for $1 \le i \le s$, is isomorphic to a direct summand of $\mathrm{Ind}^H_{H \cap g_i U g_i^{-1}}((g_i^{-1})^* N)$ for some $g_i \notin H$. Hence $H \cap g_i U g_i^{-1} \in \mathcal{V}(N_i)$.

ii. Let

$$\mathrm{Ind}_H^G(N) = M_1' \oplus \cdots \oplus M_t'$$

be a decomposition into indecomposable $R[G]$-modules. By i. and the Krull–Remak–Schmidt Theorem 1.4.7 there must exist a $1 \le l \le t$ such that N is isomorphic to a direct summand of M_l'. We define $M' := M_l'$ and $M'' := \bigoplus_{i \neq l} M_i'$. Lemma 4.3.1.iii/iv (for (M', N) instead of (M, L_i)) implies that $\mathcal{V}_0(N) \subseteq \mathcal{V}_0(M')$ and that M' and N have a common U-source.

iii. Suppose that $N_G(U) \subseteq H$ and $\mathcal{V}_0(N_i) = \mathcal{V}_0(N)$ for some $1 \le i \le s$. Then $H \cap g_i U g_i^{-1} \in \mathcal{V}(N_i) = \mathcal{V}(N)$ and therefore

$$U \subseteq h\big(H \cap g_i U g_i^{-1}\big)h^{-1} = H \cap h g_i U (h g_i)^{-1} \subseteq h g_i U (h g_i)^{-1}$$

for some $h \in H$. We conclude that $hg_i \in N_G(U) \subseteq H$ and hence $g_i \in H$ which is a contradiction. If N were isomorphic to a direct summand of M'' then $N \oplus N$ would be isomorphic to a direct summand of $\mathrm{Ind}_H^G(N)$. Hence, by i., $N \cong N_i$ for some $1 \le i \le s$ which contradicts the fact that $\mathcal{V}_0(N) \neq \mathcal{V}_0(N_i)$. □

Proposition 4.3.5 *For any $V \in \mathcal{V}_0^H(M)$ such that $N_G(V) \subseteq H$ there is a unique index $1 \le j \le r$ such that $V \in \mathcal{V}_0(L_j)$, and M is isomorphic to a direct summand of $\mathrm{Ind}_H^G(L_j)$.*

Proof According to Lemma 4.3.3 we find a finitely generated indecomposable $R[H]$-module N such that

- $V \in \mathcal{V}_0(N)$ and
- M is isomorphic to a direct summand of $\mathrm{Ind}_H^G(N)$.

Lemma 4.3.4 tells us that

$$\mathrm{Ind}_H^G(N) \cong N \oplus N_1 \oplus \cdots \oplus N_s$$

with indecomposable $R[H]$-modules N_i such that

$$\mathcal{V}_0(N_i) \neq \mathcal{V}_0(N) \quad \text{for any } 1 \le i \le s.$$

But by Lemma 4.3.2 there exists a $1 \le j \le r$ such that $V \in \mathcal{V}_0(L_j)$ and hence $\mathcal{V}_0(L_j) = \mathcal{V}_0(N)$. If we now apply the Krull–Remak–Schmidt Theorem 1.4.7 to the fact that

$$L_1 \oplus \cdots \oplus L_r \text{ is isomorphic to a direct summand of } N \oplus N_1 \oplus \cdots \oplus N_s$$

then we see that, for $i \neq j$, we have

$$\mathcal{V}_0(L_i) = \mathcal{V}_0(N_{i'}) \neq \mathcal{V}_0(N) \quad \text{for some } 1 \le i' \le s.$$

Hence j is unique with the property that $\mathcal{V}_0(L_j) = \mathcal{V}_0(N)$. Since $\mathcal{V}_0(L_j) = \mathcal{V}_0(N) \neq \mathcal{V}_0(N_i)$ for any $1 \leq i \leq s$ we furthermore have $L_j \not\cong N_1, \ldots, N_s$ and therefore necessarily $L_j \cong N$. This shows that M is isomorphic to a direct summand of $\mathrm{Ind}_H^G(L_j)$. □

In terms of the map v_H^M the Proposition 4.3.5 says the following. If the H-orbit $\mathcal{V} \subseteq \mathcal{V}_0^H(M)$ has the property that $N_G(V) \subseteq H$ for one (or any) $V \in \mathcal{V}$ then there is a unique isomorphism class $\{L\} \in \mathrm{IND}_H^0(\{M\})$ such that $\mathcal{V}_H^M(\{L\}) = \mathcal{V}$, and in fact $\{L\} \in \mathrm{IND}_H^1(\{M\})$.

We now shift our point of view and fix a subgroup $V \subseteq G$ such that $N_G(V) \subseteq H$. Then Proposition 4.3.5 gives the existence of a well-defined map

	isomorphism classes of finitely		isomorphism classes of finitely
Γ:	generated indecomposable	$\longrightarrow$	generated indecomposable
	$R[G]$-modules with vertex V		$R[H]$-modules with vertex V

where the image $\{L\} = \Gamma(\{M\})$ is characterized by the condition that L is isomorphic to a direct summand of M.

Theorem 4.3.6 (Green) *The map Γ is a bijection. The image $\{L\} = \Gamma(\{M\})$ is characterized by either of the following two conditions*:

a. *L is isomorphic to a direct summand of M*;
b. *M is isomorphic to a direct summand of* $\mathrm{Ind}_H^G(L)$.

Moreover, M and L have a common V-source.

Proof Let L be any finitely generated indecomposable $R[H]$-module with vertex V. By Lemma 4.3.4.ii we find a finitely generated indecomposable $R[G]$-module M' with vertex V such that

- L is isomorphic to a direct summand of M';
- $\mathrm{Ind}_H^G(L) = M' \oplus M''$ as $R[G]$-module, and
- M' and L have a common V-source.

In particular, $\Gamma(\{M'\}) = \{L\}$ which shows the surjectivity of Γ.

Let M be any other finitely generated indecomposable $R[G]$-module with vertex V. First we consider the case that M is isomorphic to a direct summand of $\mathrm{Ind}_H^G(L)$. Suppose that $M \not\cong M'$. Then M must be isomorphic to a direct summand of M''. Let $M = X_1 \oplus \cdots \oplus X_t$ be a decomposition into indecomposable $R[H]$-modules X_i. Then Lemma 4.3.4 implies that $\mathcal{V}_0(X_i) \neq \mathcal{V}_0(L)$ for any $1 \leq i \leq t$. But by Lemma 4.3.2 there exists a $1 \leq i \leq t$ such that $V \in \mathcal{V}_0(X_i)$ which is a contradiction. It follows that $M \cong M'$, hence that L is isomorphic to a direct summand of M, and consequently that $\Gamma(\{M\}) = \{L\}$. This proves that the condition b. characterizes the map Γ.

Secondly we consider the case that $\Gamma(\{M\}) = \{L\}$, i.e. that L is isomorphic to a direct summand of M. Then M is isomorphic to a direct summand of $\mathrm{Ind}_H^G(L)$ by

Proposition 4.3.5. Hence we are in the first case and conclude that $M \cong M'$. This establishes the injectivity of Γ. □

The bijection Γ is called the *Green correspondence* for (G, V, H). We want to establish a useful additional "rigidity" property of the Green correspondence. For this we first have to extent the concept of relative projectivity to module homomorphisms.

Definition Let $\mathcal{H}$ be a fixed set of subgroups of G.

i. An $R[G]$-module M is called relatively $\mathcal{H}$-projective if $M \cong \bigoplus_{i\in I} M_i$ is isomorphic to a direct sum of $R[G]$-modules M_i each of which is relatively $R[H_i]$-projective for some $H_i \in \mathcal{H}$.
ii. An $R[G]$-module homomorphism $\alpha : M \longrightarrow N$ is called $\mathcal{H}$-projective if there exist a relatively $\mathcal{H}$-projective $R[G]$-module X and $R[G]$-module homomorphisms $\alpha_0 : M \longrightarrow X$ and $\alpha_1 : X \longrightarrow N$ such that $\alpha = \alpha_1 \circ \alpha_0$. In case $\mathcal{H} = \{H\}$ we simply say that α is H-projective.

Exercise 4.3.7

i. An $R[G]$-module M is relatively $R[H]$-projective if and only if id_M is H-projective.
ii. If the $R[G]$-module homomorphism $\alpha : M \longrightarrow N$ is $\mathcal{H}$-projective then $\gamma \circ \alpha \circ \beta$, for any $R[G]$-module homomorphisms $\beta : M' \longrightarrow M$ and $\gamma : N \longrightarrow N'$, is $\mathcal{H}$-projective as well.
iii. If $\alpha : M \longrightarrow N$ is an H-projective $R[G]$-module homomorphism between two finitely generated $R[G]$-modules M and N then the relatively $R[H]$-projective $R[G]$-module X in the above definition can be chosen to be finitely generated as well. (Hint: Replace X by $R[G] \otimes_{R[H]} X$.)

We keep fixing our subgroups $V \subseteq H \subseteq G$, and we introduce the family of subgroups

$$\mathfrak{h} := \{H \cap gVg^{-1} : g \in G \setminus H\}.$$

Lemma 4.3.8 *Let $\alpha : M \longrightarrow N$ be a V-projective $R[H]$-module homomorphism between two finitely generated $R[G]$-modules M and N; then there exists a V-projective $R[G]$-module homomorphism $\beta : M \longrightarrow N$ and an $\mathfrak{h}$-projective $R[H]$-module homomorphism $\gamma : M \longrightarrow N$ such that $\alpha = \beta + \gamma$.*

Proof By assumption we have a commutative diagram

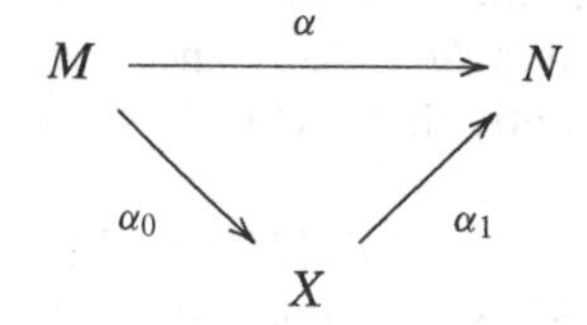

where X is a relatively $R[V]$-projective $R[H]$-module. By Exercise 4.3.7.iii we may assume that X is a finitely generated $R[H]$-module. The idea of the proof consists in the attempt to replace X by $\mathrm{Ind}_H^G(X)$. We first of all observe that, X being isomorphic to a direct summand of $\mathrm{Ind}_V^H(X)$, the $R[G]$-module $\mathrm{Ind}_H^G(X)$ is isomorphic to a direct summand of $\mathrm{Ind}_V^G(X)$ and hence is relatively $R[V]$-projective as well. We will make use of the two Frobenius reciprocities from Sect. 2.3. Let

$$X \xrightarrow{\iota} \mathrm{Ind}_H^G(X) \xrightarrow{\pi} X$$

be the $R[H]$-module homomorphisms given by

$$\iota(x)(g) := \begin{cases} g^{-1}x & \text{if } g \in H, \\ 0 & \text{if } g \notin H \end{cases}$$

and

$$\pi(\phi) := \phi(1).$$

We obviously have $\pi \circ \iota = \mathrm{id}_X$ and therefore $\mathrm{Ind}_H^G(X) = \mathrm{im}(\iota) \oplus \ker(\pi)$ as an $R[H]$-module. By applying Lemma 4.3.4.i to the indecomposable direct summands of X we see that $\ker(\pi)$ is relatively $\mathfrak{h}$-projective. But we have the commutative diagram

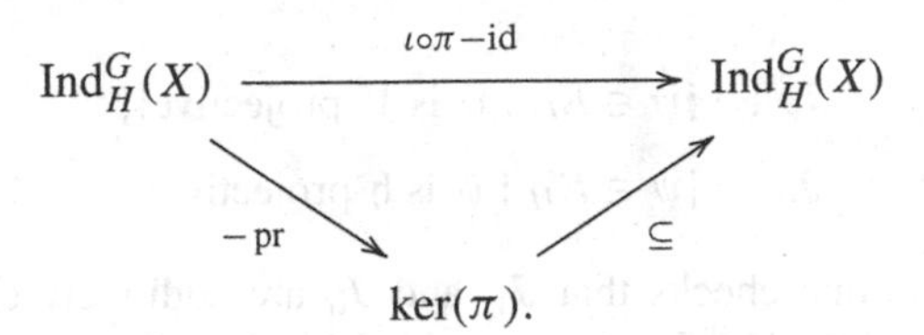

It follows that the $R[H]$-module homomorphism $\iota \circ \pi - \mathrm{id}$ is $\mathfrak{h}$-projective. By the first and second Frobenius reciprocity we have the commutative diagrams

$$\begin{array}{ccc} M & \xrightarrow{\tilde{\alpha}_0} & \mathrm{Ind}_H^G(X) \\ & \searrow^{\alpha_0} & \downarrow \pi \\ & & X \end{array} \quad \text{and} \quad \begin{array}{ccc} \mathrm{Ind}_H^G(X) & \xrightarrow{\tilde{\alpha}_1} & N, \\ \uparrow \iota & \nearrow_{\alpha_1} & \\ X & & \end{array}$$

respectively, where $\tilde{\alpha}_0$ and $\tilde{\alpha}_1$ are (uniquely determined) $R[G]$-module homomorphisms. Then $\beta := \tilde{\alpha}_1 \circ \tilde{\alpha}_0$ is a V-projective $R[G]$-module homomorphism, and $\gamma := \tilde{\alpha}_1 \circ (\iota \circ \pi - \mathrm{id}) \circ \tilde{\alpha}_0$ is a $\mathfrak{h}$-projective $R[H]$-module homomorphism (cf. Exercise 4.3.7.ii). We have

$$\begin{aligned} \alpha = \alpha_1 \circ \alpha_0 &= \tilde{\alpha}_1 \circ \tilde{\alpha}_0 + (\alpha_1 \circ \alpha_0 - \tilde{\alpha}_1 \circ \tilde{\alpha}_0) \\ &= \beta + (\tilde{\alpha}_1 \circ \iota \circ \pi \circ \tilde{\alpha}_0 - \tilde{\alpha}_1 \circ \tilde{\alpha}_0) = \beta + \tilde{\alpha}_1 \circ (\iota \circ \pi - \mathrm{id}) \circ \tilde{\alpha}_0 \\ &= \beta + \gamma. \end{aligned}$$

□

Proposition 4.3.9 *Let M be a finitely generated indecomposable $R[G]$-module with vertex V such that $N_G(V) \subseteq H \subseteq G$, and let L be its Green correspondent (i.e. $\{L\} = \Gamma(\{M\})$). For any other finitely generated $R[G]$-module M' we have:*

i. *If M' is indecomposable such that L is isomorphic to a direct summand of M' then $M' \cong M$;*
ii. *M is isomorphic to a direct summand of M' (as an $R[G]$-module) if and only if L is isomorphic to a direct summand of M' (as an $R[H]$-module).*

Proof i. By the injectivity of the Green correspondence it suffices to show that V is a vertex of M'. The assumption on M' guarantees the existence of $R[H]$-module homomorphisms $\iota : L \longrightarrow M'$ and $\pi : M' \longrightarrow L$ such that $\pi \circ \iota = \mathrm{id}_L$. Since V is a vertex of L the $R[H]$-module homomorphism $\alpha := \iota \circ \pi$ is V-projective. We therefore, by Lemma 4.3.8, may write $\alpha = \beta + \gamma$ with a V-projective $R[G]$-module endomorphism $\beta : M' \longrightarrow M'$ and an $\mathfrak{h}$-projective $R[H]$-module endomorphism $\gamma : M' \longrightarrow M'$.

Step 1: We first claim that $V \in \mathcal{V}(M')$, i.e. that $\mathrm{id}_{M'}$ is V-projective. For this we consider the inclusion of rings

$$E_G := \mathrm{End}_{R[G]}(M') \subseteq E_H := \mathrm{End}_{R[H]}(M'),$$

and we define

$$J_V := \{\psi \in E_G : \psi \text{ is } V\text{-projective}\},$$
$$J_{\mathfrak{h}} := \{\psi \in E_H : \psi \text{ is } \mathfrak{h}\text{-projective}\}.$$

Using Lemma 4.1.2 one checks that J_V and $J_{\mathfrak{h}}$ are additively closed. Moreover, Exercise 4.3.7.ii implies that J_V is a two-sided ideal in E_G and $J_{\mathfrak{h}}$ is a two-sided ideal in E_H. We have $\beta \in J_V$ and $\gamma \in J_{\mathfrak{h}}$. Furthermore

$$\alpha = \alpha^n = (\beta + \gamma)^n \equiv \beta \mod J_{\mathfrak{h}} \tag{4.3.1}$$

for any $n \in \mathbb{N}$. By Lemma 1.3.5.ii, on the other hand, E_H and hence $E_H/J_{\mathfrak{h}}$ is finitely generated as a (left) E_G-module (even as an R-module). Proposition 1.3.6.iii then implies that the E_G-module $E_H/J_{\mathfrak{h}}$ is $\mathrm{Jac}(E_G)$-adically complete. This, in particular, means that

$$\bigcap_{n \in \mathbb{N}} \big(\mathrm{Jac}(E_G)^n E_H + J_{\mathfrak{h}}\big) \subseteq J_{\mathfrak{h}}. \tag{4.3.2}$$

We now suppose that $J_V \neq E_G$. Since M' is indecomposable as an $R[G]$-module the ring E_G is local by Proposition 1.3.6 and Proposition 1.4.5. It follows that $\beta \in J_V \subseteq \mathrm{Jac}(E_G)$. We then conclude from (4.3.1) and (4.3.2) that $\alpha \in J_{\mathfrak{h}}$. Hence $\mathrm{id}_L = \pi \circ \alpha \circ \iota \in J_{\mathfrak{h}}$. We obtain that L is relatively $R[H \cap gVg^{-1}]$-projective for some $g \in G \setminus H$. But as we have seen in the proof of Lemma 4.3.4.iii the conclusion that $H \cap gVg^{-1} \in \mathcal{V}(L)$ for some $g \in G \setminus H$ is in contradiction with $V \in \mathcal{V}_0(L)$. We therefore must have $J_V = E_G$ which is the assertion that $\mathrm{id}_{M'}$ is V-projective.

Step 2: We now show that $V \in \mathcal{V}_0(M')$. Let V' be a vertex of M'. By Step 1 we have $|V'| \le |V|$, and it suffices to show that $|V'| = |V|$. But Lemma 4.3.1.i applied to M' and V' says that the order of the vertex V of the indecomposable direct summand $\iota(L)$ of M' is $\le |V'|$.

ii. The direct implication is clear since L is isomorphic to a direct summand of M. For the reverse implication let $M' = M'_1 \oplus \cdots \oplus M'_t$ be a decomposition into indecomposable $R[G]$-modules. Then L is isomorphic to a direct summand of M'_i for some $1 \le i \le t$, and the assertion i. implies that $M \cong M'_i$. □

4.4 An Example: The Group $\mathrm{SL}_2(\mathbb{F}_p)$

We fix a prime number $p \neq 2$, and we let $G := \mathrm{SL}_2(\mathbb{F}_p)$. There are the following important subgroups

$$U := \left\{ \begin{pmatrix} 1 & 0 \\ c & 1 \end{pmatrix} : c \in \mathbb{F}_p \right\} \subseteq B := \left\{ \begin{pmatrix} a & 0 \\ c & d \end{pmatrix} \in G : ad = 1 \right\}$$

of G.

Exercise

i. $[G : B] = p + 1$ and $[B : U] = p - 1$.
ii. $U \cong \mathbb{Z}/p\mathbb{Z}$ is a p-Sylow subgroup of G.
iii. $B = N_G(U)$.

A much more lengthy exercise is the following.

Exercise The group G has exactly $p + 4$ conjugacy classes which are represented by the elements:

1. $\left(\begin{smallmatrix} 1 & 0 \\ 0 & 1 \end{smallmatrix}\right), \left(\begin{smallmatrix} -1 & 0 \\ 0 & -1 \end{smallmatrix}\right)$
2. $\left(\begin{smallmatrix} 1 & 1 \\ 0 & 1 \end{smallmatrix}\right), \left(\begin{smallmatrix} -1 & 1 \\ 0 & -1 \end{smallmatrix}\right), \left(\begin{smallmatrix} 1 & \varepsilon \\ 0 & 1 \end{smallmatrix}\right), \left(\begin{smallmatrix} -1 & \varepsilon \\ 0 & -1 \end{smallmatrix}\right)$ (with $\varepsilon \in \mathbb{F}_p^\times \setminus \mathbb{F}_p^{\times 2}$ a fixed element),
3. $\left(\begin{smallmatrix} a & 0 \\ 0 & a^{-1} \end{smallmatrix}\right)$ where $a \in \mathbb{F}_p^\times \setminus \{\pm 1\}$ (up to replacing a by a^{-1}),
4. $\left(\begin{smallmatrix} 0 & -1 \\ 1 & a \end{smallmatrix}\right)$ where the polynomial $X^2 - aX + 1 \in \mathbb{F}_p[X]$ is irreducible.

The elements in 2. have order divisible by p, all the others an order prime to p.

Let k be an algebraically closed field of characteristic p.

Lemma 4.4.1 *There are exactly p isomorphism classed of simple $k[G]$-modules.*

Proof Since, by the above exercise, there are p conjugacy classes of p-regular elements in G this follows from Corollary 3.2.4. □

We want to construct explicit models for the simple $k[G]$-modules. Let $k[X,Y]$ be the polynomial ring in two variables X and Y over k. For any $g = \begin{pmatrix} a & b \\ c & d \end{pmatrix} \in G$ we define the k-algebra homomorphism

$$g\colon\ k[X,Y] \longrightarrow k[X,Y]$$
$$X \longmapsto aX + cY$$
$$Y \longmapsto bX + dY.$$

An easy explicit computation shows that in this way $k[X,Y]$ becomes a $k[G]$-module. We have the decomposition

$$k[X,Y] = \bigoplus_{n \geq 0} V_n$$

where

$$V_n := \sum_{i=0}^{n} kX^i Y^{n-i}$$

is the $k[G]$-submodule of all polynomials which are homogeneous of total degree n. We obviously have:

- $\dim_k V_n = n + 1$;
- $V_0 = k$ is the trivial $k[G]$-module;
- The G-action on $V_1 = kX + kY \cong k^2$ is the restriction to $\mathrm{SL}_2(\mathbb{F}_p)$ of the natural $\mathrm{GL}_2(k)$-action on the standard k-vector space k^2.

We will show that $V_0, V_1, \ldots, V_{p-1}$ are simple $k[G]$-modules. Since they have different k-dimensions they then must be, up to isomorphism, all simple $k[G]$-modules.

The subgroup U is generated by the element $u^+ := \begin{pmatrix} 1 & 0 \\ 1 & 1 \end{pmatrix}$. Similarly the subgroup $U^- := \{\begin{pmatrix} 1 & b \\ 0 & 1 \end{pmatrix} : b \in \mathbb{F}_p\}$ is generated by $u^- := \begin{pmatrix} 1 & 1 \\ 0 & 1 \end{pmatrix}$. In the following we fix an $n \geq 0$, and we consider in V_n the increasing sequence of vector subspaces

$$\{0\} = W_0 \subset W_1 \subset \cdots \subset W_n \subset W_{n+1} = V_n$$

defined by

$$W_i := \sum_{j=0}^{i-1} kX^j Y^{n-j}.$$

Lemma 4.4.2 *For $0 \leq i \leq n$ we have:*

i. *W_{i+1} is a $k[U]$-submodule of V_n;*
ii. *W_{i+1}/W_i is the trivial $k[U]$-module;*
iii. *if $i < p$ then each vector in $W_{i+1} \setminus W_i$ generates W_{i+1} as a $k[U]$-module.*

Proof Obviously $W_{i+1} = kX^iY^{n-i} \oplus W_i$. We have $u^+X = X + Y$ and $u^+Y = Y$ and hence

$$\begin{aligned}
u^+(X^iY^{n-i}) &= (X+Y)^iY^{n-i} = \sum_{j=0}^{i} \binom{i}{j} X^jY^{i-j}Y^{n-i} \\
&= X^iY^{n-i} + \sum_{j=0}^{i-1} \binom{i}{j} X^jY^{n-j} \\
&\equiv X^iY^{n-i} \mod W_i.
\end{aligned}$$

We now argue by induction with respect to i. The assertions i.–iii. hold trivially true for $W_1 = kY^n$. We assume that they hold true for W_i. The above congruence immediately implies i. and ii. for W_{i+1}. For iii. let $v \in W_{i+1} \setminus W_i$ be any vector. Then

$$v = aX^iY^{n-i} + w \quad \text{with } a \in k^\times \text{ and } w \in W_i.$$

Using that $u^+w - w \in W_{i-1}$ by ii. for W_i, we obtain

$$\begin{aligned}
u^+v - v &= a\bigl(u^+(X^iY^{n-i}) - X^iY^{n-i}\bigr) + (u^+w - w) \\
&\equiv aiX^{i-1}Y^{n-i+1} \mod W_{i-1}.
\end{aligned}$$

Since $i < p$ we have $ai \neq 0$ in k, and we conclude that $k[U]v$ contains the nonzero vector $u^+v - v \in W_i \setminus W_{i-1}$. Hence, by iii. for W_i, we get that $W_i \subseteq k[U]v \subseteq W_{i+1}$. Since $v \notin W_i$ we, in fact, must have $k[U]v = W_{i+1}$. □

For convenience we insert the following reminder.

Remark Let H be any finite p-group. By Proposition 2.2.7 the trivial $k[H]$-module is, up to isomorphism, the only simple $k[H]$-module. It follows that the socle of any $k[H]$-module M (cf. Lemma 1.1.5) is equal to

$$\operatorname{soc}(M) = \{x \in M : hx = x \text{ for any } h \in H\}.$$

Suppose that $M \neq \{0\}$. For any $0 \neq x \in M$ the submodule $k[H]x$ is of finite k-dimension and nonzero and therefore, by the Jordan–Hölder Proposition 1.1.2, contains a simple submodule. It follows in particular that $\operatorname{soc}(M) \neq \{0\}$.

Lemma 4.4.3 *For* $0 \le n < p$ *we have*:

i. $k[U]X^n = V_n$;
ii. kY^n *is the socle of the* $k[U]$*-module* V_n;
iii. V_n *is indecomposable as a* $k[U]$*-module.*

Proof i. This is a special cases of Lemma 4.3.2.iii. ii. The socle in question is equal to $\{v \in V_n : u^+v = v\}$. It contains $W_1 = kY^n$ by Lemma 4.4.2.ii. If $u^+v = v$ then

$k[U]v$ has k-dimension ≤ 1. On the other hand, any $0 \neq v \in V_n$ is contained in $W_{i+1} \setminus W_i$ for a unique $0 \leq i \leq n$. By Lemma 4.4.2.iii we then have $k[U]v = W_{i+1}$. Since W_{i+1} has k-dimension $i+1$ we see that, if $u^+v = v$, then necessarily $v \in W_1$.
iii. Suppose that $V_n = M \oplus N$ is the direct sum of two nonzero $k[U]$-submodules M and N. Then the socle of V_n is the direct sum of the nonzero socles of M and N and therefore has k-dimension at least 2. This contradicts ii. □

Proposition 4.4.4 *The $k[G]$-module V_n, for $0 \leq n < p$, is simple.*

Proof First of all we observe that

$$k[U^-]Y^n = V_n$$

holds true. This is proved by the same reasoning as for Lemma 4.4.3.i with only interchanging the roles of X and Y.

Let $W \subseteq V_n$ be any nonzero $k[G]$-submodule. Its socle as a $k[U]$-module is nonzero and is contained in the corresponding socle of V_n. Hence Lemma 4.4.3.ii implies that $kY^n \subseteq W$. Our initial observation then gives $V_n = k[U^-]Y^n \subseteq W$. □

The $k[U]$-module structure of V_n, for $0 \leq n < p$, can be made even more explicit. Let $k[Z]$ be the polynomial ring in one variable Z over k. Because of $(u^+ - 1)^p = u^{+p} - 1 = 0$ we have the k-algebra homomorphism

$$\begin{aligned} k[Z]/Z^p k[Z] &\longrightarrow k[U] \\ Z &\longmapsto u^+ - 1. \end{aligned}$$

Because of $(u^+)^i = ((u^+ - 1) + 1)^i$ it is surjective. But both sides have the same k-dimension p. Hence it is an isomorphism. If we view V_n, via this isomorphism, as a $k[Z]/Z^p[Z]$-module then we claim that

$$V_n \cong k[Z]/Z^{n+1}k[Z]$$

holds true. This amounts to the statement that

$$V_n \cong k[U]/(u^+ - 1)^{n+1}k[U].$$

By Lemma 4.4.3.i we have the surjective $k[U]$-module homomorphism

$$\begin{aligned} k[U] &\longrightarrow V_n \\ \sigma &\longmapsto \sigma X^n. \end{aligned}$$

On the other hand, Lemma 4.4.2.ii implies that $(u^+ - 1)W_{i+1} \subseteq W_i$ and hence by induction that $(u^+ - 1)^{n+1}X^n = 0$. We deduce that the above homomorphism induces a homomorphism

$$k[U]/(u^+ - 1)^{n+1}k[U] \longrightarrow V_n$$

which has to be an isomorphism since it is surjective and both sides have the same k-dimension $n+1$.

Remark 4.4.5 $V_{p-1} \cong k[U]$ as a $k[U]$-module.

Proof This is the case $n = p-1$ of the above discussion. □

This latter fact has an interesting consequence. For this we also need the following general properties.

Lemma 4.4.6

i. *If H is a finite p-group then any finitely generated projective $k[H]$-module is free.*
ii. *If H is a finite group and $V \subseteq H$ is a p-Sylow subgroup then the k-dimension of any finitely generated projective $k[H]$-module is divisible by $|V|$.*

Proof i. It follows from Remark 1.6.9 and Proposition 2.2.7 that $k[H]$ is a projective cover of the trivial $k[H]$-module k. Since the trivial module, up to isomorphism, is the only simple $k[H]$-module it then is a consequence of Proposition 1.7.4.i that $k[H]$, up to isomorphism, is the only finitely generated indecomposable projective $k[H]$-module. This implies the assertion by Lemma 1.1.6.

ii. Let M be a finitely generated projective $k[H]$-module. By Lemma 4.1.5 it also is projective as an $k[V]$-module. The assertion i. therefore says that $M \cong k[V]^m$ for some $m \geq 0$. It follows that $\dim_k M = m|V|$. □

Proposition 4.4.7 *Among the simple $k[G]$-modules $V_0, \ldots, V_{p-1}$ only V_{p-1} is a projective $k[G]$-module.*

Proof By Lemma 4.2.3 we have $U \in \mathcal{V}(V_{p-1})$, i.e. V_{p-1} is relatively $k[U]$-projective. But V_{p-1} also is projective as a $k[U]$-module by Remark 4.4.5. Hence it is projective as a $k[G]$-module. On the other hand, if some V_i with $0 \leq i < p$ is a projective $k[G]$-module then $i+1 = \dim_k V_i$ must be divisible by $|U| = p$ by Lemma 4.4.6.ii. It follows that $i = p-1$. □

Let M be a finitely generated indecomposable $k[G]$-module. The vertices of M, by Lemma 4.2.3, are p-groups. It follows that either $\{1\}$ or U is a vertex of M. We have observed earlier that $\{1\}$ is a vertex of M if and only if M is a projective $k[G]$-module. We postpone the investigation of this case. In the other case we may apply the Green correspondence Γ with $B = N_G(U)$ to obtain a bijection between the isomorphism classes of finitely generated indecomposable nonprojective $k[G]$-modules and the isomorphism classes of finitely generated indecomposable nonprojective $k[B]$-modules.

Example $V_0, \ldots, V_{p-2}$ of course are indecomposable $k[G]$-modules which, by Proposition 4.4.7, are nonprojective. As a consequence of Lemma 4.4.3.iii they are indecomposable as $k[B]$-modules. Hence they are their own Green correspondents.

In the following we will determine all finitely generated indecomposable $k[B]$-modules.

Another important subgroup of B is the subgroup of diagonal matrices

$$T := \left\{ \begin{pmatrix} a^{-1} & 0 \\ 0 & a \end{pmatrix} : a \in \mathbb{F}_p^\times \right\} \xrightarrow{\cong} \mathbb{F}_p^\times$$
$$\begin{pmatrix} a^{-1} & 0 \\ 0 & a \end{pmatrix} \longmapsto a.$$

Since the order $p-1$ of T is prime to p the group ring $k[T]$ is semisimple. As T is abelian and k is algebraically closed all simple $k[T]$-modules are one-dimensional. They correspond to the homomorphisms

$$\chi_i : \quad T \longrightarrow k^\times \quad \text{for } 0 \le i < p-1$$
$$\begin{pmatrix} a^{-1} & 0 \\ 0 & a \end{pmatrix} \longmapsto a^i.$$

Exercise

i. The map

$$B \xrightarrow{\text{pr}} T$$
$$\begin{pmatrix} a^{-1} & 0 \\ c & a \end{pmatrix} \longmapsto \begin{pmatrix} a^{-1} & 0 \\ 0 & a \end{pmatrix}$$

is a surjective homomorphism of groups with kernel U.

ii. $B = TU = UT$ as sets.

iii. All simple $k[B]$-modules are one-dimensional and correspond to the homomorphisms $\chi_i \circ \text{pr}$ for $0 \le i \le p-2$. (Argue that U acts trivially on any simple $k[B]$-module since in any nonzero $k[B]$-module L the subspace $\{x \in L : u^+ x = x\}$ is nonzero.)

Let S_i, for $0 \le i \le p-2$, denote the simple $k[B]$-module corresponding to $\chi_i \circ \text{pr}$.

Lemma 4.4.8 *For any $k[B]$-module L we have* $\text{rad}(L) = (u^+ - 1)L$.

Proof According to the above exercise U acts trivially on any simple $k[B]$-module and hence on $L/\text{rad}(L)$. It follows that $(u^+ - 1)L \subseteq \text{rad}(L)$. The binomial formula $(u^+)^i - 1 = ((u^+ - 1) + 1)^i - 1 = \sum_{j=1}^{i} \binom{i}{j} (u^+ - 1)^j$ implies that $(u^+ - 1)k[U] = \sum_{u \in U} (u - 1)k[U]$. Since U is normal in B it follows that $(u^+ - 1)L$ is a $k[B]$-submodule and that the U-action on $L/(u^+ - 1)L$ is trivial. Hence $k[B]$ acts on $L/(u^+ - 1)L$ through its quotient $k[T]$, and therefore $L/(u^+ - 1)L$ is a semisimple $k[B]$-module. This implies $\text{rad}(L) \subseteq (u^+ - 1)L$. □

Let $L \neq \{0\}$ be any finitely generated $k[B]$-module. According to Lemma 4.4.8 we have the sequence of $k[B]$-submodules

$$F^0(L) := L \supseteq F^1(L) := (u^+ - 1)L \supseteq \cdots \supseteq F^i(L) := (u^+ - 1)^i L \supseteq \cdots.$$

Each subquotient $F^i(L)/F^{i+1}(L)$ is a semisimple $k[B]$-module. It follows from Proposition 1.2.1 that there is a smallest $m(L) \in \mathbb{N}$ such that $F^{m(L)}(L) = \{0\}$ and $F^i(L) \neq F^{i+1}(L)$ for any $0 \leq i < L(m)$.

Our first example of a finitely generated indecomposable projective $k[B]$-module is V_{p-1} (Proposition 4.4.7, Lemmas 4.1.5, 4.4.3.iii). It follows from Lemma 4.4.2.ii that

$$F^i(V_{p-1}) = W_{p-i}, \quad \text{for } 0 \leq i \leq p, \text{ and } m(V_{p-1}) = p.$$

In particular, the sequence $F^i(V_{p-1})$ is a composition series of the $k[B]$-module V_{p-1} and $m(V_{p-1})$ is the length of V_{p-1}. For any $0 \leq i \leq p-1$ and $a \in \mathbb{F}_p^\times$ we compute

$$\begin{pmatrix} a^{-1} & 0 \\ 0 & a \end{pmatrix} (X^i Y^{p-1-i}) = (a^{-1}X)^i (aY)^{p-1-i} = a^{-2i} X^i Y^{p-1-i}.$$

This shows that

$$F^i(V_{p-1})/F^{i+1}(V_{p-1}) = W_{p-i}/W_{p-i-1} \cong S_{2i \bmod p-1} \quad \text{for } 0 \leq i \leq p-1.$$

In particular, $V_{p-1}/\mathrm{rad}(V_{p-1}) = V_{p-1}/F^1(V_{p-1}) \cong S_0$ is the trivial $k[B]$-module. Hence V_{p-1} is a projective cover of the trivial module.

It follows from Proposition 1.7.4.i that there are exactly $p-1$ isomorphism classes $\{Q_0\}, \ldots, \{Q_{p-2}\}$ of finitely generated indecomposable projective $k[B]$-modules which can be numbered in such a way that

$$Q_j/\mathrm{rad}(Q_j) = Q_j/F^1(Q_j) \cong S_j \quad \text{for } 0 \leq j < p-1.$$

Lemma 4.4.9 $Q_j \cong V_{p-1} \otimes_k S_j$ *for any* $0 \leq j < p-1$.

Proof (See Sect. 2.3 for the tensor product of two modules.) Because of Lemma 1.6.8 it suffices to show that $V_{p-1} \otimes_k S_j$ is a projective cover of S_j. But $V_{p-1} \otimes_k S_j$ obviously is finitely generated. It is indecomposable even as a $k[U]$-module. Moreover, using Lemma 4.4.8, we obtain

$$\begin{aligned} V_{p-1} \otimes_k S_j / \mathrm{rad}(V_{p-1} \otimes_k S_j) &= (V_{p-1} \otimes_k S_j)/(u^+ - 1)(V_{p-1} \otimes_k S_j) \\ &= (V_{p-1} \otimes_k S_j)/\big(((u^+ - 1)V_{p-1}) \otimes_k S_j\big) \\ &= \big(V_{p-1}/(u^+ - 1)V_{p-1}\big) \otimes_k S_j \\ &\cong S_0 \otimes_k S_j \cong S_j. \end{aligned}$$

Hence, by Remark 1.6.9, it remains to show that $V_{p-1} \otimes_k S_j$ is a projective $k[B]$-module. Let

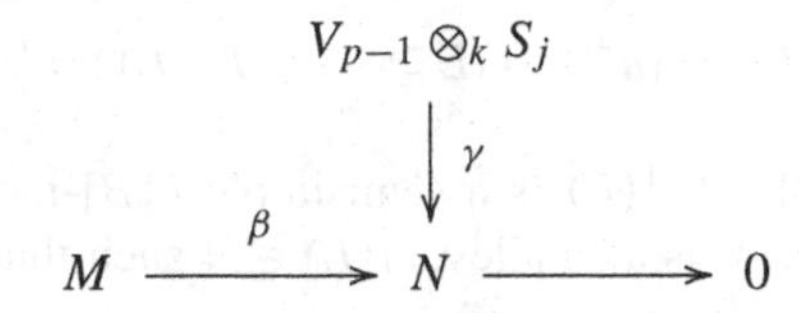

be an exact "test diagram" of $k[B]$-modules. Let $0 \leq j' < p-1$ such that $j' \equiv -j \mod p-1$. Observing that $(V_{p-1} \otimes_k S_j) \otimes_k S_{j'} = V_{p-1}$ we deduce the exact diagram

$$\begin{array}{ccccc} & & V_{p-1} & & \\ & \overset{\tilde{\alpha}}{\swarrow} & \downarrow {\scriptstyle \gamma \otimes \mathrm{id}_{S_{j'}}} & & \\ M \otimes_k S_{j'} & \underset{\beta \otimes \mathrm{id}_{S_{j'}}}{\longrightarrow} & N \otimes_k S_{j'} & \longrightarrow & 0. \end{array}$$

Since V_{p-1} is projective we find a $k[B]$-module homomorphism $\tilde{\alpha}$ such that the completed diagram commutes. But then $\alpha := \tilde{\alpha} \otimes \mathrm{id}_{S_j} : V_{p-1} \otimes_k S_j \longrightarrow (M \otimes_k S_{j'}) \otimes_k S_j = M$ satisfies $\gamma = \beta \circ \alpha$. □

Using Lemma 4.4.9 and our knowledge about V_{p-1} we deduce the following properties of the indecomposable projective $k[B]$-modules Q_j:

- $Q_j \cong k[U]$ as a $k[U]$-module.
- $F^i(Q_j)/F^{i+1}(Q_j) \cong S_{2i+j \bmod p-1}$ for $0 \leq i \leq p-1$, in particular, the $F^i(Q_j)$ form a composition series of Q_j.
- $Q_j/\mathrm{rad}(Q_j) \cong \mathrm{soc}(Q_j)$.
- The Cartan matrix of $k[B]$ is of size $(p-1) \times (p-1)$ and has the form

$$\begin{pmatrix} 3 & 0 & 2 & 0 & 2 & \ldots \\ 0 & 3 & 0 & 2 & 0 & \ldots \\ 2 & 0 & 3 & 0 & 2 & \ldots \\ 0 & 2 & 0 & 3 & 0 & \ldots \\ 2 & 0 & 2 & 0 & 3 & \ldots \\ \vdots & \vdots & \vdots & \vdots & \vdots & \end{pmatrix}.$$

In addition, the modules Q_j have the following remarkable property.

Definition Let H be a finite group. A finitely generated $k[H]$-module is called uniserial if it has a unique composition series.

Exercise In a uniserial module the members of the unique composition series are the only submodules.

Lemma 4.4.10 *Any Q_j is a uniserial $k[B]$-module.*

Proof We show the apparently stronger statement that Q_j is uniserial as a $k[U]$-module. Since $Q_j \cong k[U]$ it suffices to prove that $k[U]$ is uniserial. By our earlier discussion this amounts to showing that $k[Z]/Z^p k[Z]$ is uniserial as a module over itself. This is left as an exercise to the reader. □

It follows immediately that all

$$Q_{j,i} := Q_j/F^i(Q_j) \quad \text{for } 0 \le j < p-1 \text{ and } 0 < i \le p$$

are indecomposable $k[B]$-modules. They are pairwise nonisomorphic since any two either have different k-dimension or nonisomorphic factor modules modulo their radical. In this way we obtain $p(p-1) = |B|$ isomorphism classes of finitely generated indecomposable $k[B]$-modules.

Remark 4.4.11 Let H be a finite group and let M be a (finitely generated) $k[H]$-module. The k-linear dual $M^* := \operatorname{Hom}_k(M,k)$ is a (finitely generated) $k[H]$-module with respect to the H-action defined by

$$\begin{aligned} H \times M^* &\longrightarrow M^* \\ (h,l) &\longmapsto {}^h l(x) := l\left(h^{-1}x\right). \end{aligned}$$

Suppose that M is finitely generated. Then the map $N \longmapsto N^\perp := \{l \in M^* : l|N = 0\}$ is an inclusion reversing bijection between the set of all $k[H]$-submodules of M and the set of all $k[H]$-submodules of M^* such that

$$N^\perp = (M/N)^* \quad \text{and} \quad M^*/N^\perp = N^*.$$

It follows that M is a simple, resp. indecomposable, resp. uniserial, $k[H]$-module if and only if M^* is a simple, resp. indecomposable, resp. uniserial, $k[H]$-module.

Lemma 4.4.12 *Let H be a finite group; if the finitely generated indecomposable projective $k[H]$-modules are uniserial then all finitely generated indecomposable $k[H]$-modules are uniserial.*

Proof Let M be any finitely generated indecomposable $k[H]$-module. We consider the family of all pairs of $k[H]$-submodules $L \subset N$ of M such that N/L is nonzero and uniserial. Such pairs exist: Take, for example, any two consecutive submodules in a composition series of M. We fix a pair $L \subset N$ in this family for which the length of the factor module N/L is maximal.

In a first step we claim that there exists a $k[H]$-submodule $N_0 \subseteq N$ such that $N = N_0 \oplus L$. Since N/L is uniserial there is a unique $k[H]$-submodule $L \subseteq L_1 \subset N$ such that N/L_1 is simple. Let $\beta : P \longrightarrow N/L_1$ be a projective cover of N/L_1. By the projectivity we find a $k[H]$-module homomorphism $\alpha : P \longrightarrow N$ such that the diagram

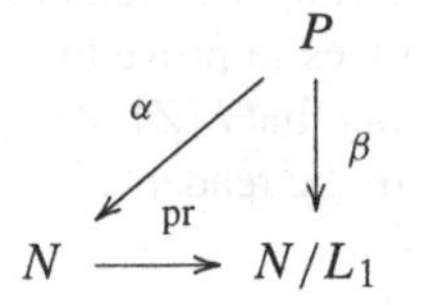

is commutative. We define $N_0 := \mathrm{im}(\alpha)$. As β is surjective we have $N_0 + L_1 = N$. If the composite map $P \xrightarrow{\alpha} N \xrightarrow{\mathrm{pr}} N/L$ were not surjective its image would be contained in L_1/L since this is the unique maximal submodule of N/L. This would imply that $N_0 \subseteq L_1$ which contradicts the above. The surjectivity of this composite map says that

$$N_0 + L = N.$$

On the other hand N_0, as a factor module of the indecomposable projective module P (cf. Proposition 1.7.4), is uniserial by assumption. Hence $\{0\} \subset N_0$ is one of the pairs of submodules in the family under consideration. Because of the surjection $N_0 \twoheadrightarrow N/L$ and the Jordan–Hölder Proposition 1.1.2 the length of N_0 cannot be smaller than the length of N/L. The maximality of the latter therefore implies that $N_0 \xrightarrow{\cong} N/L$ is an isomorphism and hence that

$$N_0 \oplus L = N.$$

The above argument, in particular, says that we may assume our pair of submodules to be of the form $\{0\} \subset N$. In the second step, we pass to the dual module M^*. The Remark 4.4.11 implies that $N^\perp \subset M^*$ is a pair of $k[H]$-submodules of M^* such that $M^*/N^\perp$ is a nonzero uniserial module of maximal possible length. Hence the argument of the first step applied to M^* leads to existence of a $k[H]$-submodule $\mathcal{M} \subseteq M^*$ such that $\mathcal{M} \oplus N^\perp = M^*$. But with M also M^* is indecomposable. We conclude that $N^\perp = \{0\}$ and consequently that $M = N$ is uniserial. □

Proposition 4.4.13 *Any finitely generated indecomposable $k[B]$-module Q is isomorphic to some $Q_{j,i}$.*

Proof It follows from Lemmas 4.4.10 and 4.4.12 that Q is uniserial. Let $\beta : Q_j \longrightarrow Q/F^1(Q)$ be a projective cover of the simple $k[B]$-module $Q/F^1(Q)$. We then find a commutative diagram of $k[B]$-module homomorphisms

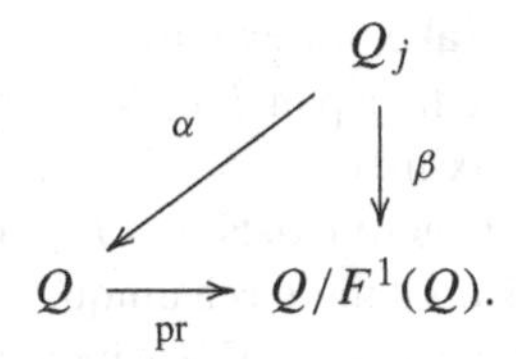

Since β is surjective the image of α cannot be contained in the unique maximal submodule $F^1(Q)$ of Q. It follows that α is surjective and induces an isomorphism $Q_{j,i} \cong Q$ for an appropriate i. □

Among the $p(p-1)$ modules $Q_{j,i}$ exactly the $p-1$ modules $Q_{j,p} = Q_j$ are projective. Hence using the Green correspondence as discussed above we derive the following result.

Proposition 4.4.14 *There are exactly $(p-1)^2$ isomorphism classes of finitely generated indecomposable nonprojective $k[G]$-modules; they are the Green correspondents of the $k[B]$-modules $Q_{j,i}$ for $0 \le j < p-1$ and $1 \le i < p$.*

Example For $0 \le n < p$ we have $V_n \cong Q_{-n \bmod p-1, n+1}$ as $k[B]$-modules.

Next we will compute the Cartan matrix of $k[G]$. But first we have to establish a useful general fact about indecomposable projective modules. For any finite group H the group ring $k[H]$ is a so-called *Frobenius algebra* in the following way. Using the k-linear form

$$\begin{aligned} \delta_1 \colon \quad k[H] &\longrightarrow k \\ \sum_{h \in H} a_h h &\longmapsto a_1 \end{aligned}$$

we introduce the k-bilinear form

$$\begin{aligned} k[H] \times k[H] &\longrightarrow k \\ (x, y) &\longmapsto \delta_1(xy). \end{aligned}$$

Remark 4.4.15

i. The above bilinear form is symmetric, i.e. $\delta_1(xy) = \delta_1(yx)$ for any $x, y \in k[H]$.
ii. The map

$$\begin{aligned} k[H] &\xrightarrow{\cong} k[H]^* = \operatorname{Hom}_k\big(k[H], k\big) \\ x &\longmapsto \delta_x(y) := \delta_1(xy) \end{aligned}$$

is a k-linear isomorphism.

Proof i. If $x = \sum_{h \in H} a_h h$ and $y = \sum_{h \in H} b_h h$ then we compute

$$\delta_1(xy) = \sum_{h \in H} a_h b_{h^{-1}} = \sum_{h \in H} b_h a_{h^{-1}} = \delta_1(yx).$$

ii. It suffices to show that the map is injective. Let $x = \sum_{h\in H} a_h h$ such that $\delta_x = 0$. For any $h_0 \in H$ we obtain

$$0 = \delta_x(h_0^{-1}) = \delta_1\Big(\sum_{h\in H} a_h h h_0^{-1}\Big) = \delta_1\Big(\sum_{h\in H} a_{hh_0} h\Big) = a_{h_0}$$

and hence $x = 0$. □

Remark 4.4.16 For any finitely generated projective $k[H]$-module P we have:

i. P^* is a projective $k[H]$-module;
ii. $P \otimes_k M$, for any $k[H]$-module M, is a projective $k[H]$-module.
iii. let M be any $k[H]$-module and let $L \subseteq N \subseteq M$ be $k[H]$-submodules such that $N/L \cong P$; then there exists a $k[H]$-submodule $M_0 \subseteq M$ such that

$$M \cong P \oplus M_0, \qquad M_0 \cap N \cong L, \quad \text{and} \quad M_0/M_0 \cap N \cong M/N.$$

Proof i. By Proposition 1.6.4 the module P is isomorphic to a direct summand of some free module $k[H]^m$. Hence P^* is isomorphic to a direct summand of $k[H]^{*m}$. Using Remark 4.4.15.ii we have the k-linear isomorphism

$$k[H] \xrightarrow{\cong} k[H]^*$$
$$\sum_{h\in H} a_h h \longmapsto \sum_{h\in H} a_h \delta_{h^{-1}}.$$

The computation

$$\delta_{(h'h)^{-1}}(y) = \delta_1(h^{-1}h'^{-1}y) = \delta_{h^{-1}}(h'^{-1}y) = {}^{h'}\delta_{h^{-1}}(y),$$

for any $h, h' \in H$ and $y \in k[H]$, shows that it is, in fact, an isomorphism of $k[H]$-modules.

ii. Again by Proposition 1.6.4 it suffices to consider the case of a free module $P = k[H]^m$. It further is enough to treat the case $m = 1$. But in the proof of Proposition 2.3.4 (applied to the subgroup $\{1\} \subseteq H$ and $W := k$, $V := M$) we have seen that

$$k[H] \otimes_k M \cong k[H]^{\dim_k M}$$

holds true.

iii. (In case $N = M$ the assertion coincides with the basic characterization of projectivity in Lemma 1.6.2.) In a first step we consider the other extreme case where $L = \{0\}$. Then N is a finitely generated projective submodule of M. Denoting by $N \xrightarrow{\iota} M$ the inclusion map we see, using i., that N^* through $\iota^* : M^* \twoheadrightarrow N^*$ is a projective factor module of M^*. Lemma 1.6.2 therefore implies the existence of a $k[H]$-module homomorphism $\sigma : N^* \longrightarrow M^*$ such that $\iota^* \circ \sigma = \mathrm{id}_{N^*}$. Dualizing again and identifying M in the usual way with a submodule of M^{**} we obtain $M = N \oplus \ker(\sigma^*|M)$.

In the general case we first use the projectivity of N/L to find a submodule $L' \subseteq N$ such that $N = L' \oplus L$ and $L' \cong N/L \cong P$. Viewing now L' as a projective submodule of M we apply the first step to obtain a submodule $M_0 \subseteq M$ such that $M = L' \oplus M_0$. Then $N = L' \oplus (M_0 \cap N)$. □

Let $\widetilde{k[H]} = \{\{P_1\}, \ldots, \{P_t\}\}$ be the set of isomorphism classes of finitely generated indecomposable projective $k[H]$-modules. It follows from Remarks 4.4.11 and 4.4.16.i that $\{P_i\} \longmapsto \{P_i^*\}$ induces a permutation of the set $\widetilde{k[H]}$. Remark 4.4.11 also implies that the $k[H]$-modules

$$\mathrm{soc}(P_i) \cong \big(P_i^* / \mathrm{rad}(P_i^*)\big)^*$$

are simple. Using Proposition 1.7.4.i we see that

$$\widehat{k[H]} = \big\{\{P_1/\mathrm{rad}(P_1)\}, \ldots, \{P_t/\mathrm{rad}(P_t)\}\big\} = \big\{\{\mathrm{soc}(P_1)\}, \ldots, \{\mathrm{soc}(P_t)\}\big\}.$$

Proposition 4.4.17 $\mathrm{soc}(P_i) \cong P_i/\mathrm{rad}(P_i)$ *for any* $1 \le i \le t$.

Proof For any $1 \le i \le t$ there is a unique index $1 \le i^* \le t$ such that $\mathrm{soc}(P_i) \cong P_{i^*}/\mathrm{rad}(P_{i^*})$. We have to show that $i^* = i$ holds true. For this we fix a decomposition

$$k[H] = N_1 \oplus \cdots \oplus N_s$$

into indecomposable (projective) submodules. Defining

$$M_i := \bigoplus_{N_j \cong P_i} N_j \quad \text{for any } 1 \le i \le t$$

we obtain a decomposition

$$k[H] = M_1 \oplus \cdots \oplus M_t.$$

We know from Corollary 1.7.5 that $M_i \ne \{0\}$ (and hence $\mathrm{soc}(M_i) \ne \{0\}$) for any $1 \le i \le t$. We also see that

$$\mathrm{soc}\big(k[H]\big) = \mathrm{soc}(M_1) \oplus \cdots \oplus \mathrm{soc}(M_t)$$

is the isotypic decomposition of the semisimple $k[H]$-module $\mathrm{soc}(k[H])$ with $\mathrm{soc}(M_i)$ being $\{\mathrm{soc}(P_i)\}$-isotypic.

Let $1 = e_1 + \cdots + e_t$ with $e_i \in M_i$. By Proposition 1.5.1 the $e_1, \ldots, e_t$ are pairwise orthogonal idempotents in $k[H]$ such that $M_i = k[H]e_i$. We claim that

$$M_i\, \mathrm{soc}(M_j) = \{0\} \quad \text{whenever } i \ne j^*.$$

Let $x \in \mathrm{soc}(M_j)$ such that $M_i x \ne \{0\}$. Since $M_i x$ is contained in the $\{\mathrm{soc}(P_j)\}$-isotypic module $\mathrm{soc}(M_j)$ it follows that $M_i x$ and *a fortiori* M_i (through $M_i \xrightarrow{\cdot x}$

$M_i x$) have a factor module isomorphic to $\mathrm{soc}(P_j) \cong P_{j^*}/\mathrm{rad}(P_{j^*})$. But $M_i/\mathrm{rad}(M_i)$ by construction is $\{P_i/\mathrm{rad}(P_i)\}$-isotypic. It follows that necessarily $i = j^*$. This shows our claim and implies that

$$e_{j^*}x = \Bigl(1 - \sum_{i \neq j^*} e_i\Bigr)x = x \quad \text{for any } x \in \mathrm{soc}(M_j).$$

Suppose that $j^* \neq j$. We then obtain, using Remark 4.4.15.i, that

$$\delta_1(x) = \delta_1(e_{j^*}x) = \delta_1(xe_{j^*}) = \delta_1(xe_je_{j^*}) = \delta_1(0) = 0$$

and therefore that

$$\{0\} = \delta_1\bigl(k[H]x\bigr) = \delta_1\bigl(xk[H]\bigr) = \delta_x\bigl(k[H]\bigr)$$

for any $x \in \mathrm{soc}(M_j)$. By Remark 4.4.15.ii this implies $\mathrm{soc}(M_j) = \{0\}$ which is a contradiction. □

Exercise (Bruhat Decomposition) $G = B \cup BwU = B \cup UwB$ with $w := \left(\begin{smallmatrix} 0 & 1 \\ -1 & 0 \end{smallmatrix}\right)$.

Lemma 4.4.18

i. *The $k[G]$-module V_p is uniserial of length 2, and there is an exact sequence of $k[G]$-modules*

$$0 \longrightarrow V_1 \xrightarrow{\alpha} V_p \xrightarrow{\beta} V_{p-2} \longrightarrow 0.$$

ii. *The socle of V_p as a $k[U]$-module is equal to $kY^p \oplus k(X^p - XY^{p-1})$.*

Proof We define the k-linear map $\beta : V_p \longrightarrow V_{p-2}$ by

$$\beta\bigl(X^iY^{p-i}\bigr) := iX^{i-1}Y^{p-1-i} \quad \text{for } 0 \le i \le p.$$

In order to see that β is a $k[G]$-module homomorphism we have to check that $\beta(gv) = g\beta(v)$ holds true for any $g \in G$ and $v \in V_p$. By additivity it suffices to consider the vectors $v = X^iY^{p-i}$ for $0 \le i \le p$. Moreover, as a consequence of the Bruhat decomposition it also suffices to consider the group elements $g = u^+$, w and $g \in T$. All these cases are easy one line computations. Obviously β is surjective with $\ker(\beta) = kX^p + kY^p$. On the other hand we have the injective ring homomorphism $\alpha : k[X,Y] \longrightarrow k[X,Y]$ defined by

$$\alpha(X) := X^p, \qquad \alpha(Y) := Y^p, \quad \text{and} \quad \alpha|k = \mathrm{id}.$$

For $g = \left(\begin{smallmatrix} a & b \\ c & d \end{smallmatrix}\right) \in G$ we compute

$$\begin{aligned} \alpha(gX) &= \alpha(aX + cY) = aX^p + cY^p \\ &= (aX + cY)^p = (gX)^p = g\bigl(X^p\bigr) \\ &= g\alpha(X) \end{aligned}$$

and similarly $\alpha(gY) = g\alpha(Y)$. This shows that α is an endomorphism of $k[G]$-modules. It obviously restricts to an isomorphism $\alpha : V_1 \xrightarrow{\cong} \ker(\beta)$. Hence we have the exact sequence in i. Since V_1 and V_{p-2} are simple $k[G]$-modules the length of V_p is 2.

Next we prove the assertion ii. The socle of V_p (as a $k[U]$-module) contains the image under α of the socle of V_1 and is mapped by β into the socle of V_{p-2}. Using Lemma 4.4.3.ii we deduce that

$$kY^p \subseteq \mathrm{soc}(V_p) \quad \text{and} \quad \beta\big(\mathrm{soc}(V_p)\big) \subseteq kY^{p-2}.$$

In particular, $\mathrm{soc}(V_p)$ is at most two-dimensional. It also follows that

$$\mathrm{soc}(V_p) \subseteq \beta^{-1}\big(kY^{p-2}\big) = kY^p \oplus kX^p \oplus kXY^{p-1}.$$

We have

$$u^+Y^p = Y^p, \qquad u^+X^p = X^p + Y^p, \qquad u^+\big(XY^{p-1}\big) = XY^{p-1} + Y^p,$$

hence $u^+(X^p - XY^{p-1}) = X^p - XY^{p-1}$ and therefore $X^p - XY^{p-1} \in \mathrm{soc}(V_p)$.

It remains to show that V_p is uniserial. Suppose that it is not. Then $V_p = (kX^p + kY^p) \oplus N$ for some $k[G]$-submodule $N \cong V_{p-2}$. We are going to use the filtration

$$kY^p = W_1 \subset W_2 \subset \cdots \subset W_p \subset W_{p+1} = V_p \quad \text{with } V_p = W_p \oplus kX^p$$

introduced before Lemma 4.4.2. The identity

$$kY^p \oplus k\big(X^p - XY^{p-1}\big) = \mathrm{soc}(V_p) = kY^p \oplus \mathrm{soc}(N)$$

implies that the socle of N contains a vector of the form $u_0 + X^p$ with $u_0 \in W_p$. Let $v = u + aX^p$ with $u \in W_p$ and $a \in k$ be any vector in N. Then $v - a(u_0 + X^p) = u - au_0 \in W_p \cap N$. If $u - au_0 \neq 0$ then this element, by Lemma 4.4.2.iii, generates, as a $k[U]$-module, W_i for some $1 \leq i \leq p$. It would follow that $kY^p = W_1 \subseteq W_i \subseteq N$ which is a contradiction. Hence $u = au_0$, $v = a(u_0 + X^p)$, and $N \subseteq k(u_0 + X^p)$. But V_{p-2} has k-dimension at least 2, and we have arrived at a contradiction again. □

Lemma 4.4.19

i. *For any* $n \geq 1$ *there is an exact sequence of* $k[G]$-*modules*

$$0 \longrightarrow V_{n-1} \xrightarrow{\gamma} V_1 \otimes_k V_n \xrightarrow{\mu} V_{n+1} \longrightarrow 0.$$

ii. *For* $1 \leq n \leq p-2$ *we have* $V_1 \otimes_k V_n \cong V_{n-1} \oplus V_{n+1}$.

Proof i. Since G acts on $k[X,Y]$ by ring automorphisms the multiplication

$$\mu\colon \quad k[X,Y] \otimes_k k[X,Y] \longrightarrow k[X,Y]$$
$$f_1 \otimes f_2 \longmapsto f_1 f_2$$

is a homomorphism of $k[G]$-modules. It restricts to a surjective map

$$\mu: \quad V_1 \otimes_k V_n \longrightarrow V_{n+1}.$$

On the other hand we have the k-linear map

$$\begin{aligned} \gamma: \quad V_{n-1} &\longrightarrow V_1 \otimes_k V_n \\ v &\longmapsto X \otimes Yv - Y \otimes Xv. \end{aligned}$$

For $g = \left(\begin{smallmatrix} a & b \\ c & d \end{smallmatrix}\right) \in G$ we compute

$$\begin{aligned} g\gamma(v) &= g(X \otimes Yv - Y \otimes Xv) = gX \otimes gYgv - gY \otimes gXgv \\ &= (aX + cY) \otimes (bX + dY)gv - (bX + dY) \otimes (aX + cY)gv \\ &= abX \otimes Xgv + adX \otimes Ygv + cbY \otimes Xgv + cdY \otimes Ygv \\ &\quad - (baX \otimes Xgv + bcX \otimes Ygv + daY \otimes Xgv + dcY \otimes Ygv) \\ &= (ad - bc)X \otimes Ygv - (ad - bc)Y \otimes Xgv \\ &= X \otimes Ygv - Y \otimes Xgv \\ &= \gamma(gv). \end{aligned}$$

Hence γ is a $k[G]$-module homomorphism. We obviously have $\mathrm{im}(\gamma) \subseteq \ker(\mu)$. Since $V_1 \otimes_k V_n = X \otimes V_n \oplus Y \otimes V_n$ it also is clear that γ is injective. But

$$\begin{aligned} \dim_k V_{n-1} = n &= 2(n+1) - (n+2) \\ &= \dim_k V_1 \otimes_k V_n - \dim_k V_{n+1} \\ &= \dim_k \ker(\mu). \end{aligned}$$

It follows that $\mathrm{im}(\gamma) = \ker(\mu)$ which establishes the exact sequence in i.

ii. In the given range of n the modules V_{n-1} and V_{n+1} are simple and nonisomorphic. As a consequence of i. the assertion ii. therefore is equivalent to $V_1 \otimes_k V_n$ having a $k[G]$-submodule which is isomorphic to V_{n+1}, resp. to the nonvanishing of $\mathrm{Hom}_{k[G]}(V_{n+1}, V_1 \otimes_k V_n)$. For $n = p - 2$ this is clear, again by i., from the projectivity of V_{p-1}. We now argue by descending induction with respect to n. Suppose that $V_1 \otimes_k V_{n+1} \cong V_n \oplus V_{n+2}$ holds true. We observe that with V_1 also V_1^* is a simple $k[G]$-module by Remark 4.4.11. But since V_1, up to isomorphism, is the only simple two-dimensional $k[G]$-module we must have $V_1^* \cong V_1$. We also recall from linear algebra that the map

$$\begin{aligned} \mathrm{Hom}_{k[G]}(V_{n+1}, V_1 \otimes_k V_n) &\xrightarrow{\cong} \mathrm{Hom}_{k[G]}\big(V_{n+1} \otimes_k V_1^*, V_n\big) \\ A &\longmapsto \big[v \otimes l \longmapsto (l \otimes \mathrm{id}_{V_n})\big(A(v)\big)\big] \end{aligned}$$

is bijective. It follows that

$$\begin{aligned}\mathrm{Hom}_{k[G]}(V_{n+1}, V_1 \otimes_k V_n) &\cong \mathrm{Hom}_{k[G]}\big(V_{n+1} \otimes_k V_1^*, V_n\big) \\ &\cong \mathrm{Hom}_{k[G]}(V_{n+1} \otimes_k V_1, V_n) \\ &\cong \mathrm{Hom}_{k[G]}(V_n \oplus V_{n+2}, V_n) \\ &\neq \{0\}. \qquad \square\end{aligned}$$

For the remainder of this section let $\{P_0\}, \ldots, \{P_{p-1}\}$ be the isomorphism classes of finitely generated indecomposable projective $k[G]$-modules numbered in such a way that

$$P_n / \mathrm{rad}(P_n) \cong V_n \quad \text{for } 0 \leq n < p$$

(cf. Proposition 1.7.4). We already know from Proposition 4.4.7 that

$$P_{p-1} \cong V_{p-1}.$$

Proposition 4.4.20 $P_{p-2} \cong V_1 \otimes_k V_{p-1}$.

Proof With V_{p-1} also the $k[G]$-module $V_1 \otimes_k V_{p-1}$ is projective by Proposition 4.4.7 and Remark 4.4.16.ii. Let

$$V_1 \otimes_k V_{p-1} = L_1 \oplus \cdots \oplus L_s$$

be a decomposition into indecomposable projective $k[G]$-modules L_i. By Lemma 4.4.19.i we have an injective homomorphism of $k[G]$-modules $\gamma : V_{p-2} \longrightarrow V_1 \otimes_k V_{p-1}$. Obviously, not all of the composed maps $V_{p-2} \xrightarrow{\gamma} V_1 \otimes_k V_{p-1} \xrightarrow{\mathrm{pr}} L_i$ can be equal to the zero map. Since V_{p-2} is simple we therefore find a $1 \leq i \leq s$ together with an injective $k[G]$-module homomorphism $V_{p-2} \longrightarrow L_i$. But according to Proposition 4.4.17 the only indecomposable projective $k[G]$-module whose socle is isomorphic to V_{p-2} is, up to isomorphism, P_{p-2}. Hence $P_{p-2} \cong L_i$, i.e. P_{p-2} is isomorphic to a direct summand of $V_1 \otimes_k V_{p-1}$. To prove our assertion it remains to check that P_{p-2} and $V_1 \otimes_k V_{p-1}$ have the same k-dimension or rather that $\dim_k P_{p-2} \geq 2p$. As already recalled from Proposition 4.4.17 we have

$$P_{p-2} / \mathrm{rad}(P_{p-2}) \cong \mathrm{soc}(P_{p-2}) \cong V_{p-2}.$$

Since P_{p-2} is indecomposable and V_{p-2} is simple and nonprojective (cf. Proposition 4.4.7) we must have $\mathrm{soc}(P_{p-2}) \subseteq \mathrm{rad}(P_{p-2})$ (cf. Remark 4.4.22.i below). It follows that $\dim_k P_{p-2} \geq 2 \dim_k V_{p-2} = 2p - 2$. On the other hand we know from Lemma 4.4.6.ii that p divides $\dim_k P_{p-2}$. As $p \neq 2$ we conclude that $\dim_k P_{p-2} \geq 2p$. $\square$

Corollary 4.4.21 *P_{p-2} is a uniserial $k[G]$-module of length* 3 *such that* $[P_{p-2}] = 2[V_{p-2}] + [V_1]$ *in* $R(k[G])$.

Proof Since we know V_{p-2} to be isomorphic to the socle of the indecomposable $k[G]$-module P_{p-2} by Proposition 4.4.17 it follows from Proposition 4.4.20 and Lemma 4.4.19.i that $[P_{p-2}] = [V_{p-2}] + [V_p]$ and that P_{p-2} is uniserial if V_p is uniserial. It remains to invoke Lemma 4.4.18.i. □

We point out the following general facts which are used repeatedly.

Remark 4.4.22 Let H be a finite group, and let M be a finitely generated projective $k[H]$-module; we then have:

i. If M is indecomposable but not simple then $\mathrm{soc}(M) \subseteq \mathrm{rad}(M)$; if moreover, $\mathrm{rad}(M)/\mathrm{soc}(M)$ is simple then M is uniserial;
ii. suppose that M has a factor module of the form $N_1 \oplus \cdots \oplus N_r$ where the N_i are simple $k[H]$-modules; let $L_i \twoheadrightarrow N_i$ be a projective cover of N_i; then $L_1 \oplus \cdots \oplus L_r$ is isomorphic to a direct summand of M.

Proof i. By Proposition 4.4.17 we have the isomorphic simple modules $\mathrm{soc}(M) \cong M/\mathrm{rad}(M)$. If $\mathrm{soc}(M) \not\subseteq \mathrm{rad}(M)$ it would follow that the projection map $\mathrm{soc}(M) \xrightarrow{\cong} M/\mathrm{rad}(M)$ is an isomorphism. Hence $M = \mathrm{soc}(M) \oplus \mathrm{rad}(M)$ which is a contradiction. Since M is indecomposable $\mathrm{soc}(M)$ is the unique minimal nonzero and $\mathrm{rad}(M)$ the unique maximal proper submodule of M. The additional assumption on M therefore guarantees that there are no other nonzero proper submodules.

ii. Let

$$M = M_1 \oplus \cdots \oplus M_s$$

be a decomposition into indecomposable projective $k[H]$-modules M_i. Then $N_1 \oplus \cdots \oplus N_r$ is a factor module of the semisimple $k[H]$-module $M/\mathrm{rad}(M) = M_1/\mathrm{rad}(M_1) \oplus \cdots \oplus M_s/\mathrm{rad}(M_s)$. The Jordan–Hölder Proposition 1.1.2 now implies the existence of an injective map $\sigma : \{1, \ldots, r\} \hookrightarrow \{1, \ldots, s\}$ such that

$$N_i \cong M_{\sigma(i)}/\mathrm{rad}(M_{\sigma(i)})$$

and hence $L_i \cong M_{\sigma(i)}$ for any $1 \le i \le r$. □

Proposition 4.4.23 *For $p > 3$ we have:*

i. $V_1 \otimes_k P_{p-2} \cong P_{p-3} \oplus V_{p-1} \oplus V_{p-1}$;
ii. $V_1 \otimes_k P_n \cong P_{n-1} \oplus P_{n+1}$ *for any* $2 \le n \le p-3$;
iii. $[P_n] = 2[V_n] + [V_{p-1-n}] + [V_{p-3-n}]$ *in* $R(k[G])$ *for any* $1 \le n \le p-3$;
iv. $V_1 \otimes_k P_1 \cong P_0 \oplus P_2 \oplus V_{p-1}$;
v. P_0 *is a uniserial* $k[G]$*-module of length* 3 *such that* $[P_0] = 2[V_0] + [V_{p-3}]$.

If $p = 3$ *then* $V_1 \otimes_k P_1 \cong P_0 \oplus V_2 \oplus V_2 \oplus V_2$ *and* $[P_0] = 3[V_0]$.

Proof We know from Corollary 4.4.21 that

$$P_{p-2}/\mathrm{rad}(P_{p-2}) \cong V_{p-2}, \qquad \mathrm{rad}(P_{p-2})/\mathrm{soc}(P_{p-2}) \cong V_1,$$
$$\mathrm{soc}(P_{p-2}) \cong V_{p-2}.$$

Hence $V_1 \otimes_k P_{p-2}$ has submodules $N \supset L$ such that

$$\begin{aligned} V_1 \otimes_k P_{p-2}/N &\cong V_1 \otimes_k V_{p-2} \cong V_{p-3} \oplus V_{p-1}, \\ N/L &\cong V_1 \otimes_k V_1 \cong V_0 \oplus V_2, \\ L &\cong V_1 \otimes_k V_{p-2} \cong V_{p-3} \oplus V_{p-1}, \end{aligned}$$

where the second column of isomorphisms comes from Lemma 4.4.19.ii. Since V_{p-1} is projective we may apply Remark 4.4.16.iii iteratively to obtain that $V_1 \otimes_k P_{p-2}$ has a factor module which is isomorphic to $V_{p-3} \oplus V_{p-1} \oplus V_{p-1}$, resp. $V_0 \oplus V_2 \oplus V_2 \oplus V_2$ if $p = 3$. Using Remark 4.4.22.i we then see that the module $P_{p-3} \oplus V_{p-1} \oplus V_{p-1}$, resp. $P_0 \oplus V_2 \oplus V_2 \oplus V_2$ if $p = 3$, is isomorphic to a direct summand of $V_1 \otimes_k P_{p-2}$. It remains to compare k-dimensions. We have $\dim_k V_1 \otimes_k P_{p-2} = 4p$. Hence we have to show that $\dim_k P_{p-3} \geq 2p$ if $p > 3$, resp. $\dim_k P_0 \geq 3$ if $p = 3$. From Propositions 4.4.17 and 4.4.7 we know (cf. Remark 4.4.22.i) that

$$\dim_k P_{p-3} \geq 2 \dim_k V_{p-3} = 2p - 4$$

and from Lemma 4.4.6.ii that p divides $\dim_k P_{p-3}$. Both together obviously imply the asserted inequalities. This establish the assertion i. as well as the first half of the case $p = 3$. We also obtain

$$\begin{aligned} [P_{p-3}] &= [V_1 \otimes_k P_{p-2}] - 2[V_{p-1}] \\ &= [V_{p-3}] + [V_{p-1}] + [V_0] + [V_2] + [V_{p-3}] + [V_{p-1}] - 2[V_{p-1}] \\ &= 2[V_{p-3}] + [V_2] + [V_0] \end{aligned} \tag{4.4.1}$$

if $p > 3$, resp.

$$[P_0] = [V_1 \otimes_k P_1] - 3[V_2] = 3[V_0]$$

if $p = 3$. This is the case $n = p - 3$ of assertion iii. and the second half of the case $p = 3$.

Next we establish the case $n = p - 3$ of assertion ii. (in particular $p \geq 5$). Using (4.4.1) and Lemma 4.4.19.ii we obtain

$$\begin{aligned} [V_1 \otimes_k P_{p-3}] &= 2[V_1 \otimes_k V_{p-3}] + [V_1 \otimes_k V_0] + [V_1 \otimes_k V_2] \\ &= 2[V_{p-4}] + 2[V_{p-2}] + [V_1] + [V_1] + [V_3] \\ &= 2[V_{p-2}] + 2[V_{p-4}] + [V_3] + 2[V_1]. \end{aligned}$$

On the other hand $V_1 \otimes_k P_{p-3}$ has the factor module $V_1 \otimes_k V_{p-3} \cong V_{p-4} \oplus V_{p-2}$. Hence Remark 4.4.22.ii says that $P_{p-4} \oplus P_{p-2}$ is isomorphic to a direct summand of $V_1 \otimes_k P_{p-3}$. By Corollary 4.4.21 the summand P_{p-2} contributes $2[V_{p-2}] + [V_1]$ to the above class, whereas the summand P_{p-4} contributes at least

$[P_{p-4}/\operatorname{rad}(P_{p-4})] + [\operatorname{soc}(P_{p-4})] = 2[V_{p-4}]$. The difference is $[V_1] + [V_3]$ which cannot come from other indecomposable projective summands P_m of $V_1 \otimes_k P_{p-3}$ since each of those would contribute another $2[V_m]$. It follows that

$$V_1 \otimes_k P_{p-3} \cong P_{p-4} \oplus P_{p-2} \quad \text{and} \quad [P_{p-4}] = 2[V_{p-4}] + [V_3] + [V_1].$$

This allows us to establish ii. and iii. by descending induction. Suppose that

$$[P_n] = 2[V_n] + [V_{p-1-n}] + [V_{p-3-n}] \quad \text{and}$$
$$[P_{n+1}] = 2[V_{n+1}] + [V_{p-2-n}] + [V_{p-4-n}]$$

hold true for some $2 \leq n \leq p-4$ (the case $n = p-4$ having been settled above). Using Lemma 4.4.19.ii we obtain on the one hand that

$$\begin{aligned}[V_1 \otimes_k P_n] &= 2[V_1 \otimes_k V_n] + [V_1 \otimes_k V_{p-1-n}] + [V_1 \otimes_k V_{p-3-n}] \\ &= 2[V_{n-1}] + 2[V_{n+1}] + [V_{p-2-n}] + [V_{p-n}] + [V_{p-4-n}] + [V_{p-2-n}] \\ &= 2[V_{n+1}] + 2[V_{n-1}] + [V_{p-n}] + 2[V_{p-2-n}] + [V_{p-4-n}].\end{aligned}$$

On the other hand $V_1 \otimes_k P_n$ has the factor module $V_1 \otimes_k V_n \cong V_{n-1} \oplus V_{n+1}$ and hence, by Remark 4.4.22.ii, a direct summand isomorphic to $P_{n-1} \oplus P_{n+1}$. This summand contributes to the above class at least $2[V_{n-1}] + 2[V_{n+1}] + [V_{p-2-n}] + [V_{p-4-n}]$. Again the difference $[V_{p-n}] + [V_{p-2-n}]$ cannot arise from other indecomposable direct summands of $V_1 \otimes_k P_n$. It therefore follows that

$$V_1 \otimes_k P_n \cong P_{n-1} \oplus P_{n+1} \quad \text{and} \quad [P_{n-1}] = 2[V_{n-1}] + [V_{p-n}] + [V_{p-2-n}].$$

It remains to prove the assertions iv. and v. Using iii. for $n = 1$ and Lemma 4.4.19.ii we have

$$\begin{aligned}[V_1 \otimes_k P_1] &= 2[V_1 \otimes_k V_1] + [V_1 \otimes_k V_{p-2}] + [V_1 \otimes_k V_{p-4}] \\ &= 2[V_0] + 2[V_2] + [V_{p-3}] + [V_{p-1}] + [V_{p-5}] + [V_{p-3}] \\ &= [V_{p-1}] + 2[V_{p-3}] + [V_{p-5}] + 2[V_2] + 2[V_0].\end{aligned}$$

In particular, $V_1 \otimes_k P_1$ has a subquotient isomorphic to the projective module V_{p-1}. It also has a factor module isomorphic to $V_1 \otimes_k V_1 \cong V_0 \oplus V_2$. Using Remarks 4.4.16.iii and 4.4.22.ii we see first that $V_1 \otimes_k P_1$ has a factor module isomorphic to $V_0 \oplus V_2 \oplus V_{p-1}$ and then that it has a direct summand isomorphic to $P_0 \oplus P_2 \oplus V_{p-1}$. By iii. for $n = 2$ the latter contributes to the above class at least $2[V_0] + 2[V_2] + [V_{p-3}] + [V_{p-5}] + [V_{p-1}]$. For a third time we argue that the difference $[V_{p-3}]$ cannot come from another indecomposable direct summand of

$V_1 \otimes_k P_1$ so that we must have

$$V_1 \otimes_k P_1 \cong P_0 \oplus P_2 \oplus V_{p-1} \quad \text{and} \quad [P_0] = 2[V_0] + [V_{p-3}].$$

It finally follows from Remark 4.4.22.i that P_0 is uniserial (for all $p \geq 3$). □

Remark 1

i. $\dim_k P_0 = \dim_k P_{p-1} = p$.
ii. $\dim_k P_n = 2p$ for $1 \leq n \leq p-2$.
iii. A closer inspection of the above proof shows that for any of the modules $P = P_1, \ldots, P_{p-3}$ the subquotient $\mathrm{rad}(P)/\mathrm{soc}(P)$ is semisimple of length 2 so that P is not uniserial.

Exercise $V_1 \otimes_k P_0 = P_1$.

We deduce from Proposition 4.4.7, Corollary 4.4.21, and Proposition 4.4.23 that the Cartan matrix of $k[G]$, which has size $p \times p$, is of the form

$$\begin{pmatrix} 3 & 0 & 0 \\ 0 & 3 & 0 \\ 0 & 0 & 1 \end{pmatrix} \quad \text{for } p = 3$$

and

$$\begin{pmatrix}
2 & & & & & & & & & & & & 1 & & 0 \\
& 2 & & & & & & & & & & \cdot & & 1 & \\
& & 2 & & & & & & & & \cdot & & 1 & & \\
& & & \cdot & & & & & & \cdot & & \cdot & & & \\
& & & & \cdot & & & & \cdot & & \cdot & & & & \\
& & & & & 2 & & 1 & & \cdot & & & & & \\
& & & & & & 3 & & 1 & & & & & & \\
& & & & & 1 & & 3 & & & & & & & \\
& & & & \cdot & & 1 & & 2 & & & & & & \\
& & & \cdot & & \cdot & & & & \cdot & & & & & \\
& & \cdot & & \cdot & & & & & & \cdot & & & & \\
& \cdot & & \cdot & & & & & & & & \cdot & & & \\
1 & & 1 & & & & & & & & & & 2 & & \\
& 1 & & & & & & & & & & & & 2 & \\
0 & & & & & & & & & & & & & & 1
\end{pmatrix} \quad \text{for } p > 3.$$

If we reorder the simple modules V_n into the sequence

$$V_0, V_{(p-1)-2}, V_2, V_{(p-1)-4}, V_4, \ldots, V_{(p-1)-1}, V_1, V_{(p-1)-3}, V_3, \ldots, V_{p-1}$$

and correspondingly the P_n then the Cartan matrix becomes block diagonal, with 3 blocks, of the form

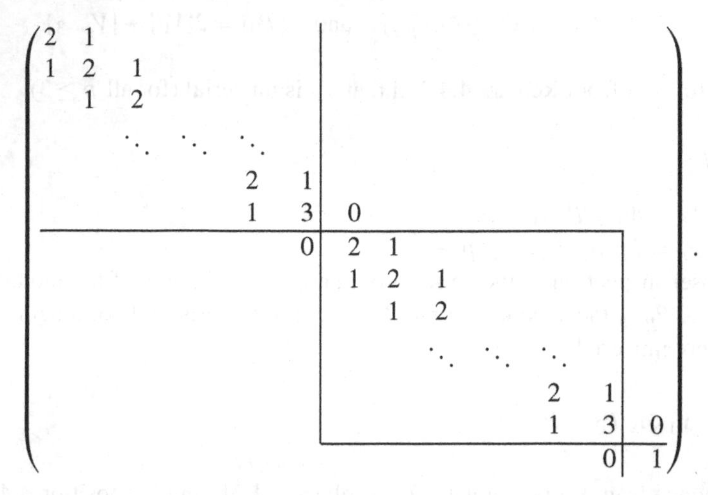

This reflects the fact, as we will see later on, that $k[G]$ has exactly three different blocks.

4.5 Green's Indecomposability Theorem

We fix a prime number p and an algebraically closed field of characteristic p. Before we come to the subject of this section we recall two facts about the matrix algebras $M_{n\times n}(k)$.

Lemma 4.5.1 *Any automorphism $T : M_{n\times n}(k) \longrightarrow M_{n\times n}(k)$ of the k-algebra $M_{n\times n}(k)$, for $n \geq 1$, is inner, i.e. there exists a matrix $T_0 \in \mathrm{GL}_n(k)$ such that $T(A) = T_0 A T_0^{-1}$ for any $A \in M_{n\times n}(k)$.*

Proof The algebra $M_{n\times n}(k)$ is simple and semisimple. Its, up to isomorphism, unique simple module is k^n with the natural action. We now use T to define a new module structure on k^n by

$$\begin{aligned} M_{n\times n}(k) \times k^n &\longrightarrow k^n \\ (A, v) &\longmapsto T(A)v \end{aligned}$$

which we denote by T^*k^n. Since any submodule of T^*k^n also is a submodule of k^n we obtain that T^*k^n is simple as well. Hence there must exist a module isomorphism $k^n \xrightarrow{\cong} T^*k^n$. This means that we find a matrix $T_0 \in \mathrm{GL}_n(k)$ such that

$$T_0 A v = T(A) T_0 v \quad \text{for any } v \in k^n \text{ and } A \in M_{n\times n}(k).$$

It follows that $T_0 A = T(A) T_0$ for any $A \in M_{n\times n}(k)$. □

Lemma 4.5.2 *Let T be an automorphism of the k-algebra $M_{p\times p}(k)$ with the property that*

$$T(e_i) = e_{i+1} \quad \text{for } 1 \le i \le p-1 \quad \text{and} \quad T(e_p) = e_1,$$

where $e_i \in M_{p\times p}(k)$ denotes the diagonal matrix with a 1 for the ith entry of the diagonal and zeros elsewhere. Then the subalgebra

$$Q := \{A \in M_{p\times p}(k) : T(A) = A\}$$

is local with $Q/\operatorname{Jac}(Q) = k$.

Proof According to Lemma 4.5.1 we find a matrix $T_0 \in \mathrm{GL}_p(k)$ such that $T(A) = T_0 A T_0^{-1}$ for any $A \in M_{p\times p}(k)$. In particular

$$T_0 e_i = e_{i+1} T_0 \quad \text{for } 1 \le i \le p-1 \quad \text{and} \quad T_0 e_p = e_1 T_0.$$

This forces T_0 to be of the form

$$\begin{pmatrix} 0 & \cdots & \cdots & \cdots & t_p \\ t_1 & 0 & & & \vdots \\ \vdots & t_2 & \ddots & & \vdots \\ \vdots & & \ddots & 0 & \vdots \\ 0 & \cdots & \cdots & t_{p-1} & 0 \end{pmatrix} \quad \text{with } t_1, \ldots, t_p \in k^\times,$$

which is conjugate to

$$\begin{pmatrix} 0 & \cdots & \cdots & \cdots & t^p \\ 1 & 0 & & & \vdots \\ \vdots & 1 & \ddots & & \vdots \\ \vdots & & \ddots & 0 & \vdots \\ 0 & \cdots & \cdots & 1 & 0 \end{pmatrix} \quad \text{where } t^p = t_1 \cdot \cdots \cdot t_p.$$

The minimal polynomial of this latter matrix is $X^p - t^p = (X-t)^p$. It follows that the Jordan normal form of T_0 is

$$\begin{pmatrix} t & & & & 0 \\ 1 & t & & & \\ & \ddots & \ddots & & \\ & & \ddots & t & \\ 0 & & & 1 & t \end{pmatrix},$$

and hence that the algebra Q is isomorphic to the algebra

$$\widetilde{Q} := \left\{ A \in M_{p\times p}(k) : \begin{pmatrix} t & & & & \\ 1 & t & & & \\ & \cdot & \cdot & & \\ & & \cdot & t & \\ & & & 1 & t \end{pmatrix} A = A \begin{pmatrix} t & & & & \\ 1 & t & & & \\ & \cdot & \cdot & & \\ & & \cdot & t & \\ & & & 1 & t \end{pmatrix} \right\}.$$

We leave it to the reader to check that:

– $\widetilde{Q}$ is the subalgebra of all matrices of the form

$$\begin{pmatrix} a & 0 & & & 0 \\ * & a & & & \\ & & \cdot & & \\ & & & a & 0 \\ * & & & * & a \end{pmatrix}.$$

–

$$I := \left\{ \begin{pmatrix} 0 & & & 0 \\ * & 0 & & \\ & & \cdot & \\ & & & 0 & \\ * & & & * & 0 \end{pmatrix} \right\}$$

is a nilpotent two-sided ideal in $\widetilde{Q}$.

– $\widetilde{Q}/I = k$.

In particular, $\widetilde{Q}$ is local with $I = \mathrm{Jac}(\widetilde{Q})$ (cf. Proposition 1.2.1.v and Proposition 1.4.1). □

We now let R be a noetherian, complete, local commutative ring such that $R/\mathrm{Jac}(R) = k$ (e.g., $R = k$ or R a $(0, p)$-ring for k), and we let G be a finite group. We fix a normal subgroup $N \subseteq G$ as well as a finitely generated indecomposable $R[N]$-module L. In this situation Mackey's Proposition 4.2.4 simplifies as follows. Let $\{g_1, \ldots, g_m\} \subseteq G$ be a set of representatives for the cosets in $N\backslash G/N = G/N$. We identify L with the $R[N]$-submodule $1 \otimes L \subseteq R[G] \otimes_{R[N]} L = \mathrm{Ind}_N^G(L)$. Then

$$\mathrm{Ind}_N^G(L) = g_1 L \oplus \cdots \oplus g_m L \quad \text{and} \quad g_i L \cong \left(g_i^{-1}\right)^* L \tag{4.5.1}$$

as $R[N]$-modules (cf. Remark 2.5.2.i). With L any $g_i L$ is an indecomposable $R[N]$-module (cf. Remark 2.5.2). Hence (4.5.1) is, in fact, a decomposition of $\mathrm{Ind}_N^G(L)$ into indecomposable $R[N]$-modules. In slight generalization of the discussion at the beginning of Sect. 2.5 we have that

$$I_G(L) := \left\{ g \in G : g^* L \cong L \text{ as } R[N]\text{-modules} \right\}$$

is a subgroup of G which contains N.

Lemma 4.5.3 *If $I_G(L) = N$ then the $R[G]$-module $\mathrm{Ind}_N^G(L)$ is indecomposable.*

Proof Suppose that the $R[G]$-module $\mathrm{Ind}_N^G(L) = M_1 \oplus M_2$ is the direct sum of two submodules. By the Krull–Remak–Schmidt Theorem 1.4.7 one of this summands, say M_1, has a direct summand M_0, as an $R[N]$-module, which is isomorphic to L. Then $g_i L \cong (g_i^{-1})^* L \cong (g_i^{-1})^* M_0 \cong g_i M_0$ is (isomorphic to) a direct summand of $g_i M_1 = M_1$. Using again Theorem 1.4.7 we see that in any decomposition of M_1 into indecomposable $R[N]$-modules all the $g_i L$ must occur up to isomorphism. On the other hand our assumption means that the $R[N]$-modules $g_i L$ are pairwise nonisomorphic, so that any $g_i L$ occurs in $\mathrm{Ind}_N^G(L)$ with multiplicity one. This shows that necessarily $M_2 = \{0\}$. □

Remark 4.5.4 $\mathrm{End}_{R[N]}(L)$ is a local ring with

$$\mathrm{End}_{R[N]}(L)/\mathrm{Jac}\big(\mathrm{End}_{R[N]}(L)\big) = k.$$

Proof The first half of the assertion is a consequence of Propositions 1.3.6 and 1.4.5. From Lemma 1.3.5.ii/iii we know that $\mathrm{End}_{R[N]}(L)$ is finitely generated as an R-module and that

$$\mathrm{Jac}(R)\,\mathrm{End}_{R[N]}(L) \subseteq \mathrm{Jac}\big(\mathrm{End}_{R[N]}(L)\big).$$

It follows that the skew field $D := \mathrm{End}_{R[N]}(L)/\mathrm{Jac}(\mathrm{End}_{R[N]}(L))$ is a finite-dimensional vector space over $R/\mathrm{Jac}(R) = k$. Since k is algebraically closed we must have $D = k$. □

Lemma 4.5.5 *Suppose that $[G : N] = p$ and that $I_G(L) = G$; then the ring $E_G := \mathrm{End}_{R[G]}(\mathrm{Ind}_N^G(L))$ is local with $E_G/\mathrm{Jac}(E_G) = k$.*

Proof We have the inclusions of rings

$$\begin{aligned} E_G = \mathrm{End}_{R[G]}\big(\mathrm{Ind}_N^G(L)\big) \subseteq E_N &:= \mathrm{End}_{R[N]}\big(\mathrm{Ind}_N^G(L)\big) \\ &\subseteq E := \mathrm{End}_R\big(\mathrm{Ind}_N^G(L)\big). \end{aligned}$$

Fixing an element $h \in G$ such that hN generates the cyclic group G/N we note that the action of h on $\mathrm{Ind}_N^G(L)$ is an R-linear automorphism and therefore defines a unit $\theta \in E^\times$. We obviously have

$$E_G = \{\alpha \in E_N : \alpha\theta = \theta\alpha\}.$$

We claim that the ring automorphism

$$\begin{aligned} \Theta\colon\ E &\longrightarrow E \\ \alpha &\longmapsto \theta\alpha\theta^{-1} \end{aligned}$$

satisfies $\Theta(E_N) = E_N$ and therefore restricts to a ring automorphism $\Theta : E_N \longrightarrow E_N$. Let $\alpha \in E_N$, which means that $\alpha \in E$ satisfies $\alpha(gx) = g\alpha(x)$ for any $g \in N$ and $x \in \mathrm{Ind}_N^G(L)$. Then

$$\begin{aligned}\theta\alpha\theta^{-1}(gx) &= h\alpha\big(h^{-1}gx\big) = h\alpha\big(h^{-1}ghh^{-1}x\big)\\ &= hh^{-1}gh\alpha\big(h^{-1}x\big) = gh\alpha\big(h^{-1}x\big)\\ &= g\big(\theta\alpha\theta^{-1}\big)(x)\end{aligned}$$

for any $g \in N$ and $x \in \mathrm{Ind}_N^G(L)$. Hence $\theta\alpha\theta^{-1} \in E_N$, and similarly $\theta^{-1}\alpha\theta \in E_N$. This proves our claim, and we deduce that

$$E_G = \big\{\alpha \in E_N : \Theta(\alpha) = \alpha\big\}.$$

An immediate consequence of this identity is the fact that

$$E_G^\times = E_G \cap E_N^\times.$$

Using Proposition 1.2.1.iv it follows that

$$\begin{aligned}E_G \cap \mathrm{Jac}(E_N) &= \big\{\alpha \in E_G : 1 + E_N\alpha \subseteq E_N^\times\big\}\\ &\subseteq \big\{\alpha \in E_G : 1 + E_G\alpha \subseteq E_N^\times \cap E_G\big\}\\ &= \big\{\alpha \in E_G : 1 + E_G\alpha \subseteq E_G^\times\big\}\\ &= \mathrm{Jac}(E_G).\end{aligned}$$

In order to compute E_N we note that $\{1, h, \ldots, h^{p-1}\}$ is a set of representatives for the cosets in G/N. The decomposition (4.5.1) becomes

$$\mathrm{Ind}_N^G(L) = L \oplus hL \oplus \cdots \oplus h^{p-1}L.$$

Furthermore, by our assumption that $I_G(L) = G$ we have $hL \cong (h^{-1})^*L \cong L$ and hence

$$\mathrm{Ind}_N^G(L) \cong L \oplus \cdots \oplus L$$

as $R[N]$-modules with p summands L on the right-hand side. We fix such an isomorphism. It induces a ring isomorphism

$$\begin{aligned}E_N &\xrightarrow{\cong} M_{p\times p}\big(E_N^0\big) \qquad \text{with } E_N^0 := \mathrm{End}_{R[N]}(L)\\ \alpha &\longmapsto \begin{pmatrix}\alpha_{11} & \cdots & \alpha_{1p}\\ \vdots & & \vdots\\ \alpha_{p1} & \cdots & \alpha_{pp}\end{pmatrix}\end{aligned}$$

where, if $x \in \operatorname{Ind}_N^G(L)$ corresponds to $(x_1, \ldots, x_p) \in L \oplus \cdots \oplus L$, then $\alpha(x)$ corresponds to

$$\left(\sum_{j=1}^{p} \alpha_{1j}(x_j), \ldots, \sum_{j=1}^{p} \alpha_{pj}(x_j)\right).$$

The automorphism Θ of E_N then corresponds to an automorphism, which we denote by T, of $M_{p\times p}(E_N^0)$. On the other hand, using Lemma 1.2.5 and Remark 4.5.4, we have

$$E_N/\operatorname{Jac}(E_N) \xrightarrow{\cong} M_{p\times p}\big(E_N^0/\operatorname{Jac}\big(E_N^0\big)\big) = M_{p\times p}(k).$$

As any ring automorphism, T respects the Jacobson radical and therefore induces a ring automorphism $\overline{T}$ of $M_{p\times p}(k)$. Introducing the subring

$$Q := \big\{\overline{A} \in M_{p\times p}(k) : \overline{T}(\overline{A}) = \overline{A}\big\}$$

we obtain the commutative diagram of injective ring homomorphisms

$$\begin{array}{ccc} E_G/E_G \cap \operatorname{Jac}(E_N) & \hookrightarrow & Q \\ \downarrow & & \downarrow \\ E_N/\operatorname{Jac}(E_N) & \xrightarrow{\cong} & M_{p\times p}(k). \end{array}$$

We remark that $\operatorname{Jac}(R)E_N \subseteq \operatorname{Jac}(E_N)$ by Lemma 1.3.5.ii/iii. It follows that the above diagram in fact is a diagram of $R/\operatorname{Jac}(R) = k$-algebras.

Next we observe that the automorphism Θ has the following property. Let $\alpha_i \in E_N$, for $1 \le i \le p$, be the endomorphism such that

$$\alpha_i | h^j L = \begin{cases} \mathrm{id} & \text{if } j = i-1, \\ 0 & \text{otherwise.} \end{cases}$$

Since $\theta(h^j L) = h^{j+1} L$ for $0 \le j < p-1$ and $\theta(h^{p-1} L) = L$ we obtain

$$\theta\alpha_i\theta^{-1} = \alpha_{i+1} \quad \text{for } 1 \le i < p \quad \text{and} \quad \theta\alpha_p\theta^{-1} = \alpha_1$$

and hence

$$\Theta(\alpha_i) = \alpha_{i+1} \quad \text{for } 1 \le i < p \quad \text{and} \quad \Theta(\alpha_p) = \alpha_1.$$

The matrix in $M_{p\times p}(E_N^0)$ corresponding to α_i is the diagonal matrix with id_L for the ith entry of the diagonal and zeros elsewhere. It immediately follows that the k-algebra automorphism $\overline{T}$ of $M_{p\times p}(k)$ satisfies the assumption of Lemma 4.5.2. We conclude that the k-algebra Q is local with $Q/\operatorname{Jac}(Q) = k$. Let $E_G \cap \operatorname{Jac}(E_N) \subseteq J \subseteq E_G$ denote the preimage of $\operatorname{Jac}(Q)$. Since $\operatorname{Jac}(Q)$ is nilpotent by Proposition 1.2.1.vi we have $J^m \subseteq E_G \cap \operatorname{Jac}(E_N) \subseteq \operatorname{Jac}(E_G)$ for some $m \ge 1$. Using Proposition 1.2.1.v it follows that $J \subseteq \operatorname{Jac}(E_G)$. The existence of the injective k-algebra homomorphism $E_G/J \hookrightarrow Q/\mathrm{Jac}(Q) = k$ then shows that $E_G/\mathrm{Jac}(E_G) = k$ must hold true. Proposition 1.4.1 finally implies that E_G is a local ring. □

Theorem 4.5.6 (Green) *Let $N \subseteq G$ be a normal subgroup such that the index $[G : N]$ is a power of p; for any finitely generated indecomposable $R[N]$-module L the $R[G]$-module $\mathrm{Ind}_N^G(L)$ is indecomposable.*

Proof Since G/N is a p-group we find a sequence of normal subgroups

$$N = N_0 \subset N_1 \subset \cdots \subset N_l = G$$

in G such that $[N_i : N_{i-1}] = p$ for any $1 \leq i \leq l$. By induction we therefore may assume that $[G : N] = p$. If $I_G(L) = N$ the assertion follows from Lemma 4.5.3. Suppose therefore that $I_G(L) = G$. Then the ring $E_G := \mathrm{End}_{R[G]}(\mathrm{Ind}_N^G(L))$ is local with $E_G/\mathrm{Jac}(E_G) = k$ by Lemma 4.5.5. It also is complete by Proposition 1.3.6.iii. Hence Proposition 1.5.11 says that 1 is the only idempotent in E_G. On the other hand, the projection of $\mathrm{Ind}_N^G(L)$ onto any nonzero direct summand as an $R[G]$-module is obviously an idempotent in E_G. It follows that $\mathrm{Ind}_N^G(L)$ must be indecomposable. □

Chapter 5
Blocks

Throughout this chapter we fix a prime number p, an algebraically closed field k of characteristic p, as well as a finite group G. Let $E := E(G) := \{e_1, \ldots, e_r\}$ be the set of all primitive idempotents in the center $Z(k[G])$ of the group ring $k[G]$. We know from Proposition 1.5.5.iii/iv that the e_i are pairwise orthogonal and satisfy $e_1 + \cdots + e_r = 1$. We recall that the e_i-block of $k[G]$ consists of all $k[G]$-modules M such that $e_i M = M$. An arbitrary $k[G]$-module M decomposes uniquely and naturally into submodules

$$M = e_1 M \oplus \cdots \oplus e_r M$$

where $e_i M$ belongs to the e_i-block. In particular, we have:

- If a module M belongs to a block then any submodule and any factor module of M belongs to the same block.
- Any indecomposable module belongs to a unique block.

By Proposition 1.5.3 the block decomposition

$$k[G] = k[G]e_1 \oplus \cdots \oplus k[G]e_r$$

of $k[G]$ is a decomposition into two-sided ideals. By Corollary 1.5.4 it is the finest such decomposition in the sense that no $k[G]e_i$ can be written as the direct sum of two nonzero two-sided ideals.

5.1 Blocks and Simple Modules

We define an equivalence relation on the set $\widetilde{k[G]}$ as follows. For $\{P\}, \{Q\} \in \widetilde{k[G]}$ we let $\{P\} \sim \{Q\}$ if there exists a sequence $\{P_0\}, \ldots, \{P_s\}$ in $\widetilde{k[G]}$ such that $P_0 \cong P$, $P_s \cong Q$, and

$$\mathrm{Hom}_{k[G]}(P_{i-1}, P_i) \neq \{0\} \quad \text{or} \quad \mathrm{Hom}_{k[G]}(P_i, P_{i-1}) \neq \{0\}$$

P. Schneider, *Modular Representation Theory of Finite Groups*,
DOI 10.1007/978-1-4471-4832-6_5, © Springer-Verlag London 2013

for any $1 \leq i \leq s$. We immediately observe that, if $\{P\}$ and $\{Q\}$ lie in different equivalence classes, then

$$\mathrm{Hom}_{k[G]}(P, Q) = \mathrm{Hom}_{k[G]}(Q, P) = \{0\}.$$

Let

$$k[G] = Q_1 \oplus \cdots \oplus Q_m$$

be a decomposition into indecomposable (projective) submodules. For any equivalence class $\mathcal{C} \subseteq \widetilde{k[G]}$ we define

$$P_{\mathcal{C}} := \bigoplus_{\{Q_i\} \in \mathcal{C}} Q_i,$$

and we obtain the decomposition

$$k[G] = \bigoplus_{\mathcal{C} \subseteq \widetilde{k[G]}} P_{\mathcal{C}}.$$

Lemma 5.1.1 *For any equivalence class $\mathcal{C} \subseteq \widetilde{k[G]}$ there exists a unique $e_{\mathcal{C}} \in E$ such that $P_{\mathcal{C}} = k[G]e_{\mathcal{C}}$; the map*

$$\text{set of equivalence classes in } \widetilde{k[G]} \xrightarrow{\sim} E$$
$$\mathcal{C} \longmapsto e_{\mathcal{C}}$$

is bijective.

Proof First let $\{P\}, \{Q\} \in \widetilde{k[G]}$ such that there exists a nonzero $k[G]$-module homomorphism $f : P \longrightarrow Q$. Suppose that P and Q belong to the e- and e'-block, respectively. Then $P/\ker(f)$ belongs to the e-block and $\mathrm{im}(f)$ to the e'-block. Since $P/\ker(f) \cong \mathrm{im}(f) \neq \{0\}$ we must have $e = e'$. This easily implies that for any equivalence class $\mathcal{C}$ there exists a unique $e_{\mathcal{C}} \in E$ such that $P_{\mathcal{C}}$ belongs to the $e_{\mathcal{C}}$-block.

Secondly, as already pointed out we have $\mathrm{Hom}_{k[G]}(P_{\mathcal{C}}, P_{\mathcal{C}'}) = \{0\}$ for any two equivalence classes $\mathcal{C} \neq \mathcal{C}'$. This implies that any $k[G]$-module endomorphism $f : k[G] \longrightarrow k[G]$ satisfies $f(P_{\mathcal{C}}) \subseteq P_{\mathcal{C}}$ for any $\mathcal{C}$. Since multiplication from the right by any element in $k[G]$ is such an endomorphism it follows that each $P_{\mathcal{C}}$ is a two-sided ideal of $k[G]$. Hence, for any $e \in E$,

$$k[G]e = \bigoplus_{\mathcal{C} \subseteq \widetilde{k[G]}} eP_{\mathcal{C}}$$

is a decomposition into two-sided ideals. It follows that there is a unique equivalence class $\mathcal{C}(e)$ such that $eP_{\mathcal{C}(e)} \neq \{0\}$; then, in fact, $k[G]e = eP_{\mathcal{C}(e)}$. Since $P_{\mathcal{C}(e)}$ belongs to the $e_{\mathcal{C}(e)}$-block we must have $e = e_{\mathcal{C}(e)}$ and $k[G]e_{\mathcal{C}(e)} = P_{\mathcal{C}(e)}$. On the other hand, given any equivalence class $\mathcal{C}$ we have $P_{\mathcal{C}} \neq \{0\}$ by Corollary 1.7.5. Hence $P_{\mathcal{C}} = e_{\mathcal{C}}P_{\mathcal{C}}$ implies $\mathcal{C}(e_{\mathcal{C}}) = \mathcal{C}$. □

Remark 5.1.2 For any $\{P\}, \{Q\} \in \widetilde{k[G]}$ we have $\operatorname{Hom}_{k[G]}(P, Q) \neq \{0\}$ if and only if the simple module $P/\operatorname{rad}(P)$ is isomorphic to a subquotient in some composition series of Q.

Proof First we suppose that there exists a nonzero $k[G]$-module homomorphism $f : P \longrightarrow Q$. The kernel of f then must be contained in the unique maximal submodule $\operatorname{rad}(P)$ of P. Hence $P/\operatorname{rad}(P) \xrightarrow[f]{\cong} f(P)/f(\operatorname{rad}(P))$.

Vice versa, let us suppose that there are submodules $L \subset N \subseteq Q$ such that $N/L \cong P/\operatorname{rad}(P)$. We then have a surjective $k[G]$-module homomorphism $P \longrightarrow N/L$ and therefore, by the projectivity of P, a commutative diagram of $k[G]$-module homomorphisms

$$\begin{array}{ccc} & & P \\ & \swarrow^{f} & \downarrow \\ N & \xrightarrow[\text{pr}]{} & N/L, \end{array}$$

where f cannot be the zero map. It follows that $\operatorname{Hom}_{k[G]}(P, Q)$ contains the nonzero composite $P \xrightarrow{f} N \xrightarrow{\subseteq} Q$. $\square$

Proposition 5.1.3 *Let $\mathcal{C} \subseteq \widetilde{k[G]}$ be an equivalence class; for any simple $k[G]$-module V the following conditions are equivalent:*

i. *V belongs to the $e_{\mathcal{C}}$-block;*
ii. *V is isomorphic to $P/\operatorname{rad}(P)$ for some $\{P\} \in \mathcal{C}$;*
iii. *V is isomorphic to a subquotient of Q for some $\{Q\} \in \mathcal{C}$.*

Proof i. $\Longrightarrow$ ii. Let V belong to the $e_{\mathcal{C}}$-block. Picking a nonzero vector $v \in V$ we obtain the surjective $k[G]$-module homomorphism

$$\begin{aligned} k[G] &\longrightarrow V \\ x &\longmapsto xv. \end{aligned}$$

It restricts to a surjective $k[G]$-module homomorphism $k[G]e_{\mathcal{C}} \longrightarrow e_{\mathcal{C}}V = V$. Lemma 5.1.1 says that $k[G]e_{\mathcal{C}} = P_{\mathcal{C}}$ is a direct sum of modules P such that $\{P\} \in \mathcal{C}$. Hence V must be isomorphic to a factor module of one of these P.

ii. $\Longrightarrow$ iii. This is trivial. iii. $\Longrightarrow$ i. By assumption V belongs to the same block as some Q with $\{Q\} \in \mathcal{C}$. But Lemma 5.1.1 says that Q, and hence V, belongs to the $e_{\mathcal{C}}$-block. $\square$

Corollary 5.1.4 *If the simple $k[G]$-module V is projective then it is, up to isomorphism, the only simple module in its block.*

Proof We consider any $\{Q\} \in \widetilde{k[G]}$ different from $\{V\}$. By Remark 4.4.16.iii the projective module V cannot be isomorphic to a subquotient of Q. Hence $\mathrm{Hom}_{k[G]}(Q, V) = \mathrm{Hom}_{k[G]}(V, Q) = \{0\}$. It follows that the equivalence class of $\{V\}$ consists of $\{V\}$ alone. □

Let us consider again the example of the group $G = \mathrm{SL}_2(\mathbb{F}_p)$ for $p > 2$. The simple $k[G]$-modules $V_0, V_1, \ldots, V_{p-1}$, as constructed in Proposition 4.4.4, are distinguished by their k-dimension $\dim_k V_n = n + 1$. Let $P_n \longrightarrow V_n$ be a projective cover. In Proposition 4.4.7, Corollary 4.4.21, and Proposition 4.4.23.iii/v we have determined the simple subquotients of each P_n. By using Remark 5.1.2 in order to translate the existence of certain subquotients into the existence of certain nonzero $k[G]$-module homomorphisms between the P_n we deduce the following facts:

- All simple subquotients of P_n for n even, resp. odd, are odd-, resp. even-, dimensional. This implies that $\mathrm{Hom}_{k[G]}(P_n, P_m) = \{0\}$ whenever n and m have different parity.
- V_{p-1} is projective.
- There exist nonzero $k[G]$-module homomorphisms:

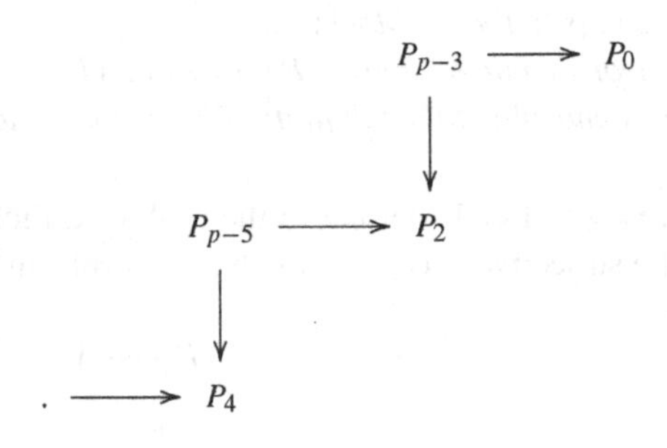

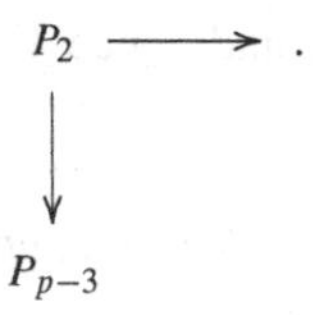

and

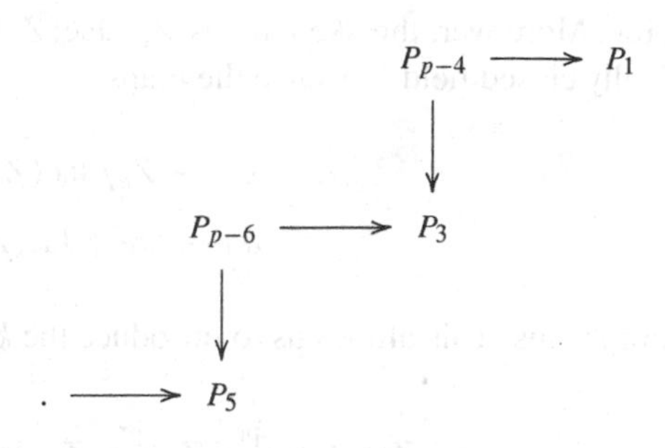

$$. \qquad\qquad .$$

$$\begin{array}{ccc} P_1 & \longrightarrow & . \\ \big\downarrow & & \\ P_{p-2} & & \end{array}$$

Together they imply that the equivalence classes in $\widetilde{k[G]}$ are

$$\big\{\{P_0\}, \{P_2\}, \ldots, \{P_{p-3}\}\big\}, \qquad \big\{\{P_1\}, \{P_3\}, \ldots, \{P_{p-2}\}\big\}, \quad \text{and} \quad \big\{\{V_{p-1}\}\big\}.$$

We conclude that $\mathrm{SL}_2(\mathbb{F}_p)$ has three blocks. More precisely, we have

$$E = \{e_{\text{even}}, e_{\text{odd}}, e_{\text{proj}}\}$$

such that a simple module V lies in the

$$\begin{aligned} e_{\text{even}}\text{-block} \quad &\Longleftrightarrow \quad \dim_k V \text{ is even}, \\ e_{\text{odd}}\text{-block} \quad &\Longleftrightarrow \quad \dim_k V \text{ is odd and} \neq p, \\ e_{\text{proj}}\text{-block} \quad &\Longleftrightarrow \quad V \cong V_{p-1}. \end{aligned}$$

5.2 Central Characters

We have the decomposition

$$Z\big(k[G]\big) = \prod_{e \in E} Z_e \quad \text{with } Z_e := Z\big(k[G]\big)e$$

of the center as a direct product of the rings Z_e (with unit element e). Since E is the set of all primitive idempotents in $Z(k[G])$ the rings Z_e do not contain any other idempotent besides their unit elements. Therefore, by Proposition 1.5.11, each Z_e is a local ring. Moreover, the skew fields $Z_e/\operatorname{Jac}(Z_e)$ are finite-dimensional over the algebraically closed field k. Hence the maps

$$\begin{aligned} \iota_e\colon\ k &\longrightarrow Z_e/\operatorname{Jac}(Z_e) \\ a &\longmapsto ae + \operatorname{Jac}(Z_e) \end{aligned}$$

are isomorphisms. This allows us to introduce the k-algebra homomorphisms

$$\chi_e\colon\ Z\big(k[G]\big) \xrightarrow{\text{pr}} Z_e \xrightarrow{\text{pr}} Z_e/\operatorname{Jac}(Z_e) \xrightarrow{\iota_e^{-1}} k,$$

which are called the *central characters* of $k[G]$.

Exercise The simple $Z(k[G])$-modules all are one-dimensional and correspond bijectively to the characters χ_e.

Proposition 5.2.1 *Let $e \in E$, and let V be a simple $k[G]$-module; then V belongs to the e-block if and only if $zv = \chi_e(z)v$ for any $z \in Z(k[G])$ and $v \in V$.*

Proof The equation $ev = \chi_e(e)v = 1v = v$ immediately implies $eV = V$ and therefore that V belongs to the e-block.

Since k is algebraically closed Schur's lemma implies that $\operatorname{End}_{k[G]}(V) = k\,\mathrm{id}_V$. This means that the homomorphism $Z(k[G]) \longrightarrow \operatorname{End}_{k[G]}(V)$ induced by the action of $Z(k[G])$ on V can be viewed as a k-algebra homomorphism $\chi : Z(k[G]) \longrightarrow k$ such that $zv = \chi(z)v$ for any $z \in Z(k[G])$ and $v \in V$. The Jacobson radical $\operatorname{Jac}(Z(k[G])) = \prod_e \operatorname{Jac}(Z_e)$, of course, lies in the kernel of χ. Suppose that V belongs to the e-block. Then $ev = v$ and $e'v = 0$ for $e \neq e' \in E$ and any $v \in V$. It follows that $\chi(e) = 1$ and $\chi(e') = 0$ for any $e' \neq e$, and hence that $\chi = \chi_e$. □

Let $\mathcal{O}(G)$ denote the set of conjugacy classes of G. For any conjugacy class $\mathcal{O} \subseteq G$ we define the element

$$\hat{\mathcal{O}} := \sum_{g \in \mathcal{O}} g \in k[G].$$

We recall that $\{\hat{\mathcal{O}} : \mathcal{O} \in \mathcal{O}(G)\}$ is a k-basis of the center $Z(k[G])$. The multiplication in $Z(k[G])$ is determined by the equations

$$\hat{\mathcal{O}}_1\hat{\mathcal{O}}_2 = \sum_{\mathcal{O} \in \mathcal{O}(G)} \mu(\mathcal{O}_1, \mathcal{O}_2; \mathcal{O})\hat{\mathcal{O}} \quad \text{with } \mu(\mathcal{O}_1, \mathcal{O}_2; \mathcal{O}) \in k.$$

(Of course, the $\mu(\mathcal{O}_1, \mathcal{O}_2; \mathcal{O})$ lie in the prime field $\mathbb{F}_p$.)

5.3 Defect Groups

The group $G \times G$ acts on $k[G]$ by

$$(G \times G) \times k[G] \longrightarrow k[G]$$
$$((g,h),x) \longmapsto gxh^{-1}.$$

In this way $k[G]$ becomes a $k[G \times G]$-module. The two-sided ideals in the ring $k[G]$ coincide with the $k[G \times G]$-submodules of $k[G]$. We see that the block decomposition

$$k[G] = \bigoplus_{e \in E} k[G]e$$

is a decomposition of the $k[G \times G]$-module $k[G]$ into indecomposable submodules. For any $e \in E$ we therefore may consider the set $\mathcal{V}_0(k[G]e)$ of vertices of the indecomposable $k[G \times G]$-module $k[G]e$. Of course, these vertices are subgroups of $G \times G$. But we have the following result. To formulate it we need the "diagonal" group homomorphism

$$\delta \colon \quad G \longrightarrow G \times G$$
$$g \longmapsto (g, g).$$

Proposition 5.3.1 *The $k[G \times G]$-module $k[G]$ is relatively $k[\delta(G)]$-projective.*

Proof Because of $(gxh^{-1}, 1)\delta(G) = (gx, h)\delta(G)$ the map

$$G \xrightarrow{\sim} (G \times G)/\delta(G)$$
$$x \longmapsto (x, 1)\delta(G)$$

is an isomorphism of $G \times G$-sets. By Lemma 2.4.6.i it induces an isomorphism

$$k[G] \xrightarrow{\cong} \mathrm{Ind}_{\delta(G)}^{G \times G}(k)$$

of $k[G \times G]$-modules. The assertion therefore follows from Proposition 4.1.6. □

Corollary 5.3.2 *For any $e \in E$ the indecomposable $k[G \times G]$-module $k[G]e$ has a vertex of the form $\delta(H)$ for some subgroup $H \subseteq G$; if $H' \subseteq G$ is another subgroup such that $\delta(H')$ is a vertex of $k[G]e$ then H and H' are conjugate in G.*

Proof By Proposition 5.3.1 and Lemma 4.1.2 the $k[G \times G]$-module $k[G]e$, being a direct summand of $k[G]$, is relatively $k[\delta(G)]$-projective. This proves the first half of the assertion. By Proposition 4.2.5 there exists an element $(g, h) \in G \times G$ such that $\delta(H') = (g, h)\delta(H)(g, h)^{-1}$. It follows that $H' = gHg^{-1}$. □

Definition Let $e \in E$; the subgroups $D \subseteq G$ such that $\delta(D) \in \mathcal{V}_0(k[G]e)$ are called the defect groups of the e-block.

Defect groups exist and are p-subgroups of G by Lemma 4.2.3. The defect groups of a single block form a conjugacy class of subgroups of G.

Lemma 5.3.3 *Let $e \in E$, and let D be a defect group of the e-block; then $k[G]e$ as a $k[\delta(G)]$-module is relatively $k[\delta(D)]$-projective.*

Proof Let

$$k[G]e = L_1 \oplus \cdots \oplus L_s$$

be a decomposition of the indecomposable $k[G \times G]$-module $k[G]e$ into indecomposable $k[\delta(G)]$-modules. By Lemma 4.1.2 it suffices to show that $\delta(D) \in \mathcal{V}(L_i)$ for any $1 \leq i \leq s$. According to Lemma 4.3.1.i we find elements $(g_i, h_i) \in G \times G$ such that

$$(g_i, h_i)\delta(D)(g_i, h_i)^{-1} \cap \delta(G) \in \mathcal{V}(L_i).$$

But the equation $(g_i, h_i)(d, d)(g_i, h_i)^{-1} = (g, g)$ with $d \in D$ and $g \in G$ implies $g_i d g_i^{-1} = h_i d h_i^{-1}$ and therefore

$$(g_i, h_i)\delta(d)(g_i, h_i)^{-1} = \delta(g_i)\delta(d)\delta(g_i)^{-1}.$$

It follows that

$$(g_i, h_i)\delta(D)(g_i, h_i)^{-1} \cap \delta(G) \subseteq \delta(g_i)\delta(D)\delta(g_i)^{-1}$$

and hence that $\delta(D) \in \mathcal{V}(L_i)$ (cf. Exercise 4.2.2.ii and Lemma 4.2.1). □

Proposition 5.3.4 *Let $e \in E$, and let D be a defect group of the e-block; any $k[G]$-module M belonging to the e-block is relatively $k[D]$-projective.*

Proof We denote by $(k[G]e)^{\mathrm{ad}}$ the k-vector space $k[G]e$ viewed as a $k[G]$-module through the group isomorphism $G \xrightarrow[\delta]{\cong} \delta(G)$. This means that G acts on $(k[G]e)^{\mathrm{ad}}$ by

$$\begin{aligned} G \times \left(k[G]e\right)^{\mathrm{ad}} &\longrightarrow \left(k[G]e\right)^{\mathrm{ad}} \\ (g, x) &\longmapsto gxg^{-1}. \end{aligned}$$

As a consequence of Lemma 5.3.3 the module $(k[G]e)^{\mathrm{ad}}$ is relatively $k[D]$-projective. We therefore find, by Proposition 4.1.6, a $k[D]$-module L such that $(k[G]e)^{\mathrm{ad}}$ is isomorphic to a direct summand of $\mathrm{Ind}_D^G(L)$. On the other hand we consider the k-linear maps

$$\begin{array}{ccccc} M & \xrightarrow{\alpha} & (k[G]e)^{\mathrm{ad}} \otimes_k M & \xrightarrow{\beta} & M \\ v & \longmapsto & e \otimes v & & \\ & & x \otimes v & \longmapsto & xv. \end{array}$$

Because of

$$\alpha(gv) = e \otimes gv = geg^{-1} \otimes gv = g(e \otimes v) = g\alpha(v)$$

and

$$\beta\big(g(x \otimes v)\big) = \beta\big(gxg^{-1} \otimes gv\big) = gxg^{-1}gv = gxv = g\beta(x \otimes v)$$

both maps are $k[G]$-module homomorphisms. The composite map satisfies $\beta\alpha(v) = ev = v$ and hence is the identity map. It follows that M is isomorphic to a direct summand of $(k[G]e)^{\mathrm{ad}} \otimes_k M$. Together we obtain that M is isomorphic to a direct summand of $\mathrm{Ind}_D^G(L) \otimes_k M$. But, as we have used before in the proof of Proposition 2.3.4, there are isomorphisms of $k[G]$-modules

$$\mathrm{Ind}_D^G(L) \otimes_k M \cong \big(k[G] \otimes_{k[D]} L\big) \otimes_k M \cong k[G] \otimes_{k[D]} (L \otimes_k M) \cong \mathrm{Ind}_D^G(L \otimes_k M).$$

Therefore, Proposition 4.1.6 implies that M is relatively $k[D]$-projective. □

The last result says that a defect group of an e-block contains a vertex of any finitely generated indecomposable module in this block. Later on (Proposition 5.4.7) we will see that the defect group occurs among these vertices. Hence defect groups can be characterized as being the largest such vertices.

At this point we start again from a different end. For any $x \in G$ we let $C_G(x) := \{g \in G : gx = xg\}$ denote the centralizer of x. Let $\mathcal{O} \subseteq G$ be a conjugacy class.

Definition The p-Sylow subgroups of the centralizers $C_G(x)$ for $x \in \mathcal{O}$ are called the defect groups of the conjugacy class $\mathcal{O}$.

For any p-subgroup $P \subseteq G$ we define

$$I_P := \sum \{k\hat{\mathcal{O}} : P \text{ contains a defect group of } \mathcal{O}\} \subseteq Z\big(k[G]\big).$$

Exercise 5.3.5

i. Any two defect groups of $\mathcal{O}$ are conjugate in G.
ii. If P is a p-Sylow subgroup then $I_P = Z(k[G])$.
iii. If $P \subseteq P'$ then $I_P \subseteq I_{P'}$.
iv. I_P only depends on the conjugacy class of P.

Lemma 5.3.6 *Let $\mathcal{O}_1, \mathcal{O}_2$, and $\mathcal{O}$ in $\mathcal{O}(G)$ be such that $\mu(\mathcal{O}_1, \mathcal{O}_2; \mathcal{O}) \neq 0$; if the p-subgroup $P \subseteq G$ centralizes an element of $\mathcal{O}$ then it also centralizes elements of $\mathcal{O}_1$ and $\mathcal{O}_2$.*

Proof Suppose that P centralizes $x \in \mathcal{O}$. We define

$$X := \{(y_1, y_2) \in \mathcal{O}_1 \times \mathcal{O}_2 : y_1 y_2 = x\}$$

as a P-set through

$$P \times X \longrightarrow X$$
$$(g, (y_1, y_2)) \longmapsto (g y_1 g^{-1}, g y_2 g^{-1}).$$

One checks that

$$|X| = \mu(\mathcal{O}_1, \mathcal{O}_2; \mathcal{O}) \neq 0 \quad \text{in } k.$$

On the other hand let $X = X_1 \cup \cdots \cup X_m$ be the decomposition of X into its P-orbits. Since P is a p-group the $|X_i|$ all are powers of p. Hence either $|X_i| = 0$ in k or $|X_i| = 1$. We see that we must have at least one P-orbit $Y = \{(y_1, y_2)\} \subseteq X$ consisting of one point. This means that P centralizes $y_1 \in \mathcal{O}_1$ and $y_2 \in \mathcal{O}_2$. □

Proposition 5.3.7 *Let $P_1, P_2 \subseteq G$ be p-subgroups; then*

$$I_{P_1} I_{P_2} \subseteq \sum_{g \in G} I_{P_1 \cap g P_2 g^{-1}}.$$

Proof For $i = 1, 2$ let $\mathcal{O}_i \in \mathcal{O}(G)$ such that P_i contains a defect subgroup D_i of $\mathcal{O}_i$. We have to show that

$$\hat{\mathcal{O}}_1 \hat{\mathcal{O}}_2 = \sum_{\mathcal{O} \in \mathcal{O}(G)} \mu(\mathcal{O}_1, \mathcal{O}_2; \mathcal{O}) \hat{\mathcal{O}} \in \sum_{g \in G} I_{P_1 \cap g P_2 g^{-1}}.$$

Obviously we only need to consider any $\mathcal{O} \in \mathcal{O}(G)$ such that $\mu(\mathcal{O}_1, \mathcal{O}_2; \mathcal{O}) \neq 0$. We pick a defect subgroup D of $\mathcal{O}$. By definition D centralizes an element of $\mathcal{O}$. Lemma 5.3.6 says that D also centralizes elements of $\mathcal{O}_1$ and $\mathcal{O}_2$. We therefore find elements $g_i \in G$, for $i = 1, 2$, such that

$$D \subseteq g_1 D_1 g_1^{-1} \cap g_2 D_2 g_2^{-1} \subseteq g_1 P_1 g_1^{-1} \cap g_2 P_2 g_2^{-1}.$$

Setting $g := g_1^{-1} g_2$ and using Exercise 5.3.5.iv it follows that

$$k\hat{\mathcal{O}} \subseteq I_{g_1 P_1 g_1^{-1} \cap g_2 P_2 g_2^{-1}} = I_{P_1 \cap g P_2 g^{-1}}.$$ □

Corollary 5.3.8 *I_P, for any p-subgroup $P \subseteq G$, is an ideal in $Z(k[G])$.*

Proof Let $P' \subseteq G$ be a fixed p-Sylow subgroup. Using Exercise 5.3.5.ii/iii we deduce from Proposition 5.3.7 that

$$I_P Z(k[G]) = I_P I_{P'} \subseteq \sum_{g \in G} I_{P \cap g P' g^{-1}} \subseteq I_P.$$ □

Remark 5.3.9 Let $e \in E$, and let $\mathcal{P}$ be a set of p-subgroups of G; if $e \in \sum_{P \in \mathcal{P}} I_P$ then $I_Q e = Z_e$ for some $Q \in \mathcal{P}$.

Proof We have

$$e = ee \in \sum_{P \in \mathcal{P}} I_P e \subseteq Z_e.$$

As noted at the beginning of Sect. 5.2 the ring Z_e is local. By Corollary 5.3.8 each $I_P e$ is an ideal in Z_e. Since $e \notin \mathrm{Jac}(Z_e)$ we must have $I_Q e \nsubseteq \mathrm{Jac}(Z_e)$ for at least one $Q \in \mathcal{P}$. But $I_Q e$ then contains a unit so that necessarily $I_Q e = Z_e$. □

For any subgroup $H \subseteq G$ we have in $k[G]$ the subring

$$K[G]^{\mathrm{ad}(H)} := \{x \in k[G] : hx = xh \text{ for any } h \in H\}.$$

We note that $k[G]^{\mathrm{ad}(G)} = Z(k[G])$. Moreover, there is the k-linear "trace map"

$$\begin{aligned} \mathrm{tr}_H : \quad k[G]^{\mathrm{ad}(H)} &\longrightarrow Z(k[G]) \\ x &\longmapsto \sum_{g \in G/H} gxg^{-1}. \end{aligned}$$

It satisfies

$$\mathrm{tr}_H(yx) = \mathrm{tr}_H(xy) = y\,\mathrm{tr}_H(x)$$

for any $x \in k[G]^{\mathrm{ad}(H)}$ and $y \in Z(k[G])$.

Lemma 5.3.10 $I_P \subseteq \mathrm{im}(\mathrm{tr}_P)$ *for any* p*-subgroup* $P \subseteq G$.

Proof It suffices to show that $\hat{\mathcal{O}}$ lies in the image of tr_P for any conjugacy class $\mathcal{O} \in \mathcal{O}(G)$ such that $Q := P \cap C_G(x)$ is a p-Sylow subgroup of $C_G(x)$ for some $x \in \mathcal{O}$. Since $[C_G(x) : Q]$ is prime to p and hence invertible in k we may define the element

$$y := \frac{1}{[C_G(x) : Q]} \sum_{h \in P/Q} hxh^{-1} \in k[G]^{\mathrm{ad}(P)}.$$

We compute

$$\begin{aligned} \mathrm{tr}_P(y) &= \sum_{g \in G/P} g\left(\frac{1}{[C_G(x) : Q]} \sum_{h \in P/Q} hxh^{-1}\right)g^{-1} = \frac{1}{[C_G(x) : Q]} \sum_{g \in G/Q} gxg^{-1} \\ &= \sum_{g \in G/C_G(x)} gxg^{-1} = \hat{\mathcal{O}}. \end{aligned}$$

□

Proposition 5.3.11 *Let* $P \subseteq G$ *be a* p*-subgroup, and let* $e \in I_P$ *be an idempotent; for any* $k[G]$*-module* M *the* $k[G]$*-submodule* eM *is relatively* $k[P]$*-projective.*

Proof According to Lemma 5.3.10 there exists an element $y \in k[G]^{\mathrm{ad}(P)}$ such that $e = \mathrm{tr}_P(y)$. Since y commutes with the elements in P the map

$$\psi:\ eM \longrightarrow eM$$
$$v \longmapsto yv$$

is a $k[P]$-module homomorphism. Let $\{g_1, \ldots, g_m\}$ be a set of representatives for the cosets in G/P. Then $\sum_{i=1}^m g_i y g_i^{-1} = e$, which translates into the identity

$$\sum_{i=1}^m g_i \psi g_i^{-1} = \mathrm{id}_{eM}.$$

Our assertion therefore follows from Lemma 4.1.7. □

Corollary 5.3.12 *Let $e \in E$, and let $P \subseteq G$ be a p-subgroup; if $e \in I_P$ then P contains a defect subgroup of the e-block.*

Proof By assumption we have

$$e = \sum_{i=1}^s a_i \hat{\mathcal{O}}_i$$

with $a_i \in k$ and such that $\mathcal{O}_1, \ldots, \mathcal{O}_s$ are all conjugacy classes such that $P \cap C_G(x_i)$, for some $x_i \in \mathcal{O}_i$, is a p-Sylow subgroup of $C_G(x_i)$. We define a central idempotent ε in $k[G \times G]$ as follows. Obviously

$$\mathcal{O}(G \times G) = \mathcal{O}(G) \times \mathcal{O}(G).$$

For $\mathcal{O} \in \mathcal{O}(G)$ we let $\mathcal{O}^{-1} := \{g^{-1} : g \in \mathcal{O}\} \in \mathcal{O}(G)$. We put

$$\varepsilon := \sum_{i,j=1}^s a_i a_j \big(\mathcal{O}_i, \mathcal{O}_j^{-1}\big)^\wedge \in Z\big(k[G \times G]\big).$$

With e also ε is nonzero. To compute ε^2 it is more convenient to write $e = \sum_{g \in G} a_g g$. We note that

- $a_g = 0$ if $g \notin \mathcal{O}_1 \cup \cdots \cup \mathcal{O}_s$, and
- that $e^2 = e$ implies $\sum_{g_1 g_2 = g} a_{g_1} a_{g_2} = a_g$.

Using the former we have $\varepsilon = \sum_{g,h \in G} a_g a_h (g, h^{-1})$. The computation

$$\varepsilon^2 = \Big[\sum_{g_1,h_1 \in G} a_{g_1} a_{h_1} \big(g_1, h_1^{-1}\big)\Big]\Big[\sum_{g_2,h_2 \in G} a_{g_2} a_{h_2} \big(g_2, h_2^{-1}\big)\Big]$$
$$= \sum_{g,h \in G} \Big(\sum_{g_1 g_2 = g,\, h_1 h_2 = h} a_{g_1} a_{g_2} a_{h_1} a_{h_2}\Big)\big(g, h^{-1}\big)$$

$$= \sum_{g,h\in G} \Big(\sum_{g_1g_2=g} a_{g_1}a_{g_2}\Big)\Big(\sum_{h_1h_2=h} a_{h_1}a_{h_2}\Big)(g,h^{-1})$$
$$= \sum_{g,h\in G} a_g a_h \big(g,h^{-1}\big)$$
$$= \varepsilon$$

now shows that ε indeed is an idempotent. For any pair (i,j) the group

$$(P\times P)\cap C_{G\times G}\big((x_i,x_j^{-1})\big) = \big(P\cap C_G(x_i)\big)\times\big(P\cap C_G(x_j^{-1})\big)$$
$$= \big(P\cap C_G(x_i)\big)\times\big(P\cap C_G(x_j)\big)$$

is a p-Sylow subgroup of $C_{G\times G}((x_i,x_j^{-1})) = C_G(x_i)\times C_G(x_j)$. It follows that

$$\varepsilon\in I_{P\times P}\subseteq Z\big(k[G\times G]\big).$$

Hence we may apply Proposition 5.3.11 to $P\times P\subseteq G\times G$ and ε and obtain that the $k[G\times G]$-module $\varepsilon k[G]$ is relatively $k[P\times P]$-projective. But applying $\varepsilon = \sum_{i,j} a_i a_j \sum_{x\in\mathcal{O}_i}\sum_{y\in\mathcal{O}_j}(x,y^{-1})$ to any $v\in k[G]$ gives

$$\varepsilon v = \sum_{i,j} a_i a_j \sum_{x\in\mathcal{O}_i}\sum_{y\in\mathcal{O}_j} xvy = \sum_{i,j} a_i a_j \hat{\mathcal{O}}_i v \hat{\mathcal{O}}_j$$
$$= \Big(\sum_i a_i\hat{\mathcal{O}}_i\Big) v \Big(\sum_j a_j\hat{\mathcal{O}}_j\Big)$$
$$= eve$$

and shows that

$$\varepsilon k[G] = ek[G]e = k[G]e.$$

It follows that $P\times P$ contains a subgroup of the form $(g,h)\delta(D)(g,h)^{-1}$ with $g,h\in G$ and D a defect subgroup of the e-block. We deduce that P contains the defect subgroup gDg^{-1}. □

Later on (Theorem 5.5.8.i) we will establish the converse of Corollary 5.3.12. Hence the defect groups of the e-block can also be characterized as being the smallest p-subgroups $D\subseteq G$ such that $e\in I_D$.

5.4 The Brauer Correspondence

Let $e\in E$, let D be a defect group of the e-block, and put $N := N_G(D)$. Obviously $N\times N$ contains the normalizer $N_{G\times G}(\delta(D))$. Hence, by Green's Theorem 4.3.6, the indecomposable $k[G \times G]$-module $k[G]e$ has a Green correspondent which is the, up to isomorphism, unique indecomposable direct summand with vertex $\delta(D)$ of $k[G]e$ as a $k[N \times N]$-module. Can we identify this Green correspondent?

We first establish the following auxiliary but general result.

Lemma 5.4.1 *Let P be any finite p-group and $Q \subseteq P$ be any subgroup; the $k[P]$-module $\mathrm{Ind}_Q^P(k)$ is indecomposable with vertex Q.*

Proof Using the second Frobenius reciprocity in Sect. 2.3 we obtain

$$\mathrm{Hom}_{k[P]}\big(k, \mathrm{Ind}_Q^P(k)\big) = \mathrm{Hom}_{k[Q]}(k, k) = k\,\mathrm{id}.$$

Since the trivial module k is the only simple $k[P]$-module by Proposition 2.2.7 this implies that the socle of the $k[P]$-module $\mathrm{Ind}_Q^P(k)$ is one-dimensional. But by the Jordan–Hölder Proposition 1.1.2 any nonzero $k[P]$-module has a nonzero socle. It follows that $\mathrm{Ind}_Q^P(k)$ must be indecomposable. In particular, by Proposition 4.1.6, $\mathrm{Ind}_Q^P(k)$ has a vertex V contained in Q. So, again by Proposition 4.1.6, $\mathrm{Ind}_Q^P(k)$ is isomorphic to a direct summand of

$$\begin{aligned}\mathrm{Ind}_V^P\big(\mathrm{Ind}_Q^P(k)\big) &\cong \bigoplus_{i=1}^{m} \mathrm{Ind}_V^P\big(\mathrm{Ind}_{V \cap g_i Q g_i^{-1}}^V\big((g_i^{-1})^* k\big)\big) \\ &= \bigoplus_{i=1}^{m} \mathrm{Ind}_{V \cap g_i Q g_i^{-1}}^P(k).\end{aligned}$$

Here $\{g_1, \ldots, g_m\} \subseteq G$ is a set of representatives for the double cosets in $V \backslash G / Q$, and the decomposition comes from Mackey's Proposition 4.2.4. By what we have shown already each summand is indecomposable. Hence the Krull–Remak–Schmidt Theorem 1.4.7 implies that $\mathrm{Ind}_Q^P(k) \cong \mathrm{Ind}_{V \cap g_i Q g_i^{-1}}^P(k)$ for some $1 \le i \le m$. Comparing k-dimensions we deduce that $|Q| \le |V|$ and consequently that $Q = V$. □

Remark 5.4.2 In passing we observe that any p-subgroup $Q \subseteq G$ occurs as a vertex of some finitely generated indecomposable $k[G]$-module. Let $P \subseteq G$ be a p-Sylow subgroup containing Q. Lemma 5.4.1 says that the indecomposable $k[P]$-module $\mathrm{Ind}_Q^P(k)$ has the vertex Q. It then follows from Lemma 4.3.4 that $\mathrm{Ind}_Q^G(k)$ has an indecomposable direct summand with vertex Q. Since the trivial $k[G]$-module k is a direct summand of $\mathrm{Ind}_P^G(k)$ (as a $p \nmid [G : P]$) we, in particular, see that the p-Sylow subgroups of G are the vertices of k. We further deduce, using Proposition 5.3.4, that the p-Sylow subgroups of G also are the defect groups of the block to which the trivial module k belongs.

We also need a few technical facts about the $k[G \times G]$-module $k[G]$ when we view it as a $k[H \times H]$-module for some subgroup $H \subseteq G$. For any $y \in G$ the double coset $HyH \subseteq G$ is an $H \times H$-orbit in G for the action

$$(H \times H) \times G \longrightarrow G$$

$$\big((h_1, h_2), g\big) \longmapsto h_1 g h_2^{-1},$$

and so $k[HyH]$ is a $k[H \times H]$-submodule of $k[G]$.

We remind the reader that the centralizer of a subgroup $Q \subseteq G$ is the subgroup $C_G(Q) = \{g \in G : gh = hg \text{ for any } h \in G\}$.

Lemma 5.4.3

i. $k[HyH] \cong \mathrm{Ind}^{H \times H}_{(1,y)^{-1}\delta(H \cap yHy^{-1})(1,y)}(k)$ *as* $k[H \times H]$*-modules.*

ii. *If* H *is a* p*-group then* $k[HyH]$ *is an indecomposable* $k[H \times H]$*-module with vertex* $(1, y)^{-1}\delta(H \cap yHy^{-1})(1, y)$.

iii. *Suppose that* $y \notin H$ *and that* $C_G(Q) \subseteq H$ *for some* p*-subgroup* $Q \subseteq H$*; then no indecomposable direct summand of the* $k[H \times H]$*-module* $k[HyH]$ *has a vertex containing* $\delta(Q)$.

Proof i. We observe that

$$\begin{aligned}
\big\{(h_1, h_2) \in H \times H : (h_1, h_2)y = y\big\} &= \big\{(h_1, h_2) \in H \times H : h_1 y h_2^{-1} = y\big\} \\
&= \big\{(h_1, h_2) \in H \times H : h_2 = y^{-1} h_1 y\big\} \\
&= \big\{\big(h, y^{-1}hy\big) : h \in H, y^{-1}hy \in H\big\} \\
&= \big\{(1, y)^{-1}(h, h)(1, y) : h \in H \cap yHy^{-1}\big\} \\
&= (1, y)^{-1}\delta\big(H \cap yHy^{-1}\big)(1, y).
\end{aligned}$$

Hence, by Remark 2.4.2, the map

$$\begin{aligned}
H \times H/(1, y)^{-1}\delta\big(H \cap yHy^{-1}\big)(1, y) &\xrightarrow{\cong} HyH \\
(h_1, h_2) \cdots &\longmapsto h_1 y h_2^{-1}
\end{aligned}$$

is an isomorphism of $H \times H$-sets. The assertion now follows from Lemma 2.4.6.i.

ii. This follows from i. and Lemma 5.4.1.

iii. By Proposition 4.1.6 the assertion i. implies that each indecomposable summand in question has a vertex contained in $(1, y)^{-1}\delta(H \cap yHy^{-1})(1, y)$. Suppose therefore that this latter group contains an $H \times H$-conjugate of $\delta(Q)$. We then find elements $h_1, h_2 \in H$ such that

$$(h_1, h_2)\delta(Q)(h_1, h_2)^{-1} \subseteq (1, y)^{-1}\delta\big(H \cap yHy^{-1}\big)(1, y) \subseteq (1, y)^{-1}\delta(G)(1, y),$$

hence

$$(h_1, yh_2)\delta(Q)(h_1, yh_2)^{-1} \subseteq \delta(G),$$

and therefore

$$h_1 g h_1^{-1} = yh_2 g (yh_2)^{-1} \quad \text{for any } g \in Q.$$

It follows that $h_1^{-1} y h_2 \in C_G(Q) \subseteq H$. But this implies $y \in H$ which is a contradiction. □

For any p-subgroup $D \subseteq G$ we put

$$E_D(G) := \{e \in E : D \text{ is a defect subgroup of the } e\text{-block}\}.$$

Theorem 5.4.4 (Brauer's First Main Theorem) *For any p-subgroup $D \subseteq G$ and any subgroup $H \subseteq G$ such that $N_G(D) \subseteq H$ we have a bijection*

$$B: \quad E_D(G) \xrightarrow{\sim} E_D(H)$$

such that the $k[H \times H]$-module $k[H]B(e)$ is the Green correspondent of the $k[G \times G]$-module $k[G]e$.

Proof First of all we point out that, for any two idempotents $e \neq e'$ in $E(G)$, the $k[G]$-modules $k[G]e$ and $k[G]e'$ belong to two different blocks and therefore cannot be isomorphic. *A fortiori* they cannot be isomorphic as $k[G \times G]$-modules. A corresponding statement holds, of course, for the $k[H \times H]$-modules $k[H]f$ with $f \in E(H)$.

Let $e \in E_D(G)$. Since $H \times H$ contains $N_{G \times G}(\delta(D))$ the Green correspondent $\Gamma(k[G]e)$ of the indecomposable $k[G \times G]$-module $k[G]e$ exists, is an indecomposable direct summand of $k[G]e$ as a $k[H \times H]$-module, and has the vertex $\delta(D)$ (see Theorem 4.3.6). But we have the decomposition

$$k[G] = \bigoplus_{y \in H \backslash G / H} k[HyH] \tag{5.4.1}$$

as a $k[H \times H]$-module. The Krull–Remak–Schmidt Theorem 1.4.7 implies that $\Gamma(k[G]e)$ is isomorphic to a direct summand of $k[HyH]$ for some $y \in G$. By Lemma 5.4.3.iii we must have $y \in H$. Hence $\Gamma(k[G]e)$ is isomorphic to a direct summand of the $k[H \times H]$-module $k[H]$. It follows that

$$\Gamma\big(k[G]e\big) \cong k[H]f$$

for some $f \in E_D(H)$. Our initial observation applied to H says that $B(e) := f$ is uniquely determined.

For $e \neq e'$ in $E_D(G)$ the $k[G \times G]$-modules $k[G]e$ and $k[G]e'$ are nonisomorphic. By the injectivity of the Green correspondence the $k[H \times H]$-modules $\Gamma(k[G]e)$ and $\Gamma(k[G]e')$ are nonisomorphic as well. Hence $B(e) \neq B(e')$. This shows that B is injective.

For the surjectivity of B let $f \in E_D(H)$. The decomposition (5.4.1) shows that $k[H]f$, as a $k[H \times H]$-module, is isomorphic to a direct summand of $k[G]$ and hence, by the Krull–Remak–Schmidt Theorem 1.4.7, to a direct summand of $k[G]e$ for some $e \in E(G)$. It follows from Proposition 4.3.9 that necessarily $k[H]f \cong \Gamma(k[G]e)$ and $e \in E_D(G)$. In particular, $f = B(e)$. □

The bijection B in Theorem 5.4.4 is a particular case of the *Brauer correspondence* whose existence comes from the following result.

Lemma 5.4.5 *Let $f \in E(H)$ such that $C_G(D) \subseteq H$ for one (or equivalently any) defect group D of the f-block; we then have:*

i. *There is a unique $\beta(f) := e \in E(G)$ such that $k[H]f$ is isomorphic, as a $k[H \times H]$-module, to a direct summand of $k[G]e$;*
ii. *any defect group of the f-block is contained in a defect group of the e-block.*

Proof i. The argument for the existence of e was already given at the end of the proof of Theorem 5.4.4. Suppose that $k[H]f$ also is isomorphic to a direct summand of $k[G]e'$ for a second idempotent $e \neq e' \in E(G)$. Then $k[H]f \oplus k[H]f$ is isomorphic to a direct summand of $k[G] = \bigoplus_{\tilde{e} \in E(G)} k[G]\tilde{e}$. Since $k[H]f$, but not $k[H]f \oplus k[H]f$ is isomorphic to a direct summand of $k[H]$ the decomposition (5.4.1) shows that $k[H]f$ must be isomorphic to a direct summand of $k[HyH]$ for some $y \in G \setminus H$. But this is impossible according to Lemma 5.4.3.iii (applied with $Q = D$).

ii. Let $\tilde{D}$ be a defect group of the e-block. By Lemma 4.3.1.i applied to $k[G]e \cong k[H]f \oplus \cdots$ and $H \times H \subseteq G \times G$ we find a $\gamma = (g_1, g_2) \in G \times G$ such that $\delta(D) \subseteq \gamma\delta(\tilde{D})\gamma^{-1}$. Then $D \subseteq g_1\tilde{D}g_1^{-1}$. □

For any p-subgroup $D \subseteq G$ we put

$$E_{\geq D}(G) := \{e \in E : D \text{ is contained in a defect group of the } e\text{-block}\}.$$

We fix a subgroup $H \subseteq G$ and a p-subgroup $D \subseteq H$, and we suppose that $C_G(D) \subseteq H$. Any other p-subgroup $D \subseteq D' \subseteq H$ then satisfies $C_G(D') \subseteq C_G(D) \subseteq H$ as well. Therefore, according to Lemma 5.4.5, we have the well-defined map

$$\beta: \quad E_{\geq D}(H) \longrightarrow E_{\geq D}(G).$$

Theorem 5.4.6 *Let $f \in E(H)$, and suppose that there is a finitely generated indecomposable $k[H]$-module N belonging to the f-block which has a vertex V that satisfies $C_G(V) \subseteq H$; we then have:*

i. *$C_G(D) \subseteq H$ for any defect group D of the f-block; in particular, $\beta(f)$ is defined;*
ii. *if a finitely generated indecomposable $k[G]$-module M has a direct summand, as a $k[H]$-module, isomorphic to N then M belongs to the $\beta(f)$-block.*

Proof By Proposition 5.3.4 we may assume that $V \subseteq D$. Then $C_G(D) \subseteq C_G(V) \subseteq H$ which proves i. Let $e \in E(G)$ such that M belongs to the e-block. Suppose that $e \neq \beta(f)$.

We deduce from (5.4.1), by multiplying by f, the decomposition

$$fk[G] = k[H]f \oplus fX \quad \text{with } X := \bigoplus_{y \in H\backslash G/H, y \notin H} k[HyH].$$

Since f commutes with the elements in H this is a decomposition of $k[H \times H]$-modules. Similarly we have the decomposition

$$fk[G] = fk[G]e \oplus fk[G](1-e)$$

as a $k[H \times H]$-module. So $fk[G]e$ is a direct summand of $k[H]f \oplus fX$. By assumption the indecomposable $k[H \times H]$-module $k[H]f$ is not isomorphic to a direct summand of $k[G]e = fk[G]e \oplus (1-f)k[G]e$. Hence it is not isomorphic to a direct summand of $fk[G]e$. It therefore follows from the Krull–Remak–Schmidt Theorem 1.4.7 that $fk[G]e$ must be isomorphic to a direct summand of fX and hence of $X = fX \oplus (1-f)X$ as a $k[H \times H]$-module. Using Lemma 5.4.3.iii we conclude that no indecomposable direct summand Y of the $k[H \times H]$-module $fk[G]e$ has a vertex containing $\delta(V)$. We consider any indecomposable direct summand Z of Y as a $k[\delta(H)]$-module. Lemma 4.3.1.i applied to $Y = Z \oplus \cdots$ and $\delta(H) \subseteq H \times H$ implies that any vertex of Z is contained in some vertex of Y (as a $k[H \times H]$-module). It follows that no vertex of Z contains $\delta(V)$. Using the notation introduced in the proof of Proposition 5.3.4 we conclude that no indecomposable direct summand of the $k[H]$-module $(fk[G]e)^{\mathrm{ad}}$ has a vertex containing V. If we analyze the arguments in the proof of Proposition 5.3.4 then, in the present context, they give the following facts:

- $fM = efM$ is isomorphic to a direct summand of $(fk[G]e)^{\mathrm{ad}} \otimes_k fM$.
- If Z is an indecomposable direct summand of $(fk[G]e)^{\mathrm{ad}}$ with vertex U then $Z \otimes_k fM$ is relatively $k[U]$-projective. This implies that the vertices of the indecomposable direct summands of $(fk[G]e)^{\mathrm{ad}} \otimes_k fM$ are contained in vertices of the indecomposable direct summands of $(fk[G]e)^{\mathrm{ad}}$.

We deduce that the indecomposable direct summands of fM have no vertices containing V. But $N = fN$ is a direct summand of M and hence of fM and it does have vertex V. This is a contradiction. We therefore must have $e = \beta(f)$. □

Proposition 5.4.7 *Let $e \in E(G)$ and let D be a defect group of the e-block; then there exists a finitely generated indecomposable $k[G]$-module M belonging to the e-block which has the vertex D.*

Proof Let $H = N_G(D)$ and let X be any simple $k[H]$-module in the $B(e)$-block of $k[H]$. We claim that X has the vertex D. But D is a normal p-subgroup of H. Hence as a $k[D]$-module X is a direct sum

$$X = k \oplus \cdots \oplus k$$

of trivial $k[D]$-modules k (cf. Proposition 2.2.7 and Theorem 2.5.3.i). On the other hand the $B(e)$-block has the defect group D so that, by Proposition 5.3.4, X is relatively $k[D]$-projective. We therefore may apply Lemma 4.3.1 to $D \subseteq H$ and the above decomposition of X and obtain that any vertex of the trivial $k[D]$-module also is a vertex of the $k[H]$-module X. The former, by Lemma 5.4.1, are equal to D. Hence X has the vertex D. Green's Theorem 4.3.6 now tells us that X is the

Green correspondent of some finitely generated indecomposable $k[G]$-module M with vertex D. Moreover, Theorem 5.4.6 implies that M belongs to the $\beta(B(e)) = e$-block. □

Proposition 5.4.8 *For any $e \in E(G)$ the following conditions are equivalent*:

i. *The e-block has the defect group $\{1\}$*;
ii. *any $k[G]$-module belonging to the e-block is semisimple*;
iii. *the k-algebra $k[G]e$ is semisimple*;
iv. *$k[G]e \cong M_{n\times n}(k)$ as k-algebras for some $n \geq 1$*;
v. *there is a simple $k[G]$-module X belonging to the e-block which is projective.*

Proof i. $\Longrightarrow$ ii. By Proposition 5.3.4 any module M belonging to the e-block is relatively $k[\{1\}]$-projective and hence projective. In particular, M/N is projective for any submodule $N \subseteq M$; it follows that $M \cong N \oplus M/N$. This implies that M is semisimple (cf. Proposition 1.1.4).

ii. $\Longrightarrow$ iii. $k[G]e$ as a $k[G]$-module belongs to the e-block and hence is semisimple.

iii. $\Longrightarrow$ iv. Since e is primitive in $Z(k[G])$ the semisimple k-algebra $k[G]e$ must, in fact, be simple (cf. Corollary 1.5.4). Since k is algebraically closed any finite-dimensional simple k-algebra is a matrix algebra.

iv. $\Longrightarrow$ v. The simple module of a matrix algebra is projective.

v. $\Longrightarrow$ i. We have seen in Corollary 5.1.4 that X is, up to isomorphism, the only simple module belonging to the e-block. This implies that any finitely generated indecomposable module belonging to the e-block is isomorphic to X and hence has the vertex $\{1\}$. Therefore the e-block has the defect group $\{1\}$ by Proposition 5.4.7. □

5.5 Brauer Homomorphisms

Let $D \subseteq G$ be a p-subgroup. There is the obvious k-linear map

$$\tilde{s}: \quad k[G] \longrightarrow k\big[C_G(D)\big] \subseteq k\big[N_G(D)\big]$$
$$\sum_{g\in G} a_g g \longmapsto \sum_{g\in C_G(D)} a_g g.$$

We observe that the centralizer $C_G(D)$ is a normal subgroup of the normalizer $N_G(D)$. This implies that the intersection $\mathcal{O} \cap C_G(D)$, for any conjugacy class $\mathcal{O} \in \mathcal{O}(G)$, on the one hand is contained in $C_G(D)$, of course, but on the other hand is a union of full conjugacy classes in $\mathcal{O}(N_G(D))$. We conclude that $\tilde{s}$ restricts to a k-linear map

$$s: \quad Z(k[G]) \longrightarrow Z(k[N_G(D)])$$
$$\hat{\mathcal{O}} \longmapsto \sum_{o \in \mathcal{O}(N_G(D)), o \subseteq \mathcal{O} \cap C_G(D)} \hat{o}.$$

Lemma 5.5.1 *s is a homomorphism of k-algebras.*

Proof Clearly s respects the unit element. It remains to show that s is multiplicative. Let $\mathcal{O}_1, \mathcal{O}_2 \in \mathcal{O}(G)$. We have

$$\hat{\mathcal{O}}_1 \hat{\mathcal{O}}_2 = \sum_{g \in G} |B_g| g \quad \text{with } B_g := \{(g_1, g_2) \in \mathcal{O}_1 \times \mathcal{O}_2 : g_1 g_2 = g\}$$

and hence

$$s(\hat{\mathcal{O}}_1 \hat{\mathcal{O}}_2) = \sum_{g \in C_G(D)} |B_g| g.$$

On the other hand

$$s(\hat{\mathcal{O}}_1) s(\hat{\mathcal{O}}_2) = \Big(\sum_{g_1 \in \mathcal{O}_1 \cap C_G(D)} g_1 \Big) \Big(\sum_{g_2 \in \mathcal{O}_2 \cap C_G(D)} g_2 \Big) = \sum_{g \in C_G(D)} |C_g| g$$

with

$$C_g := \{(g_1, g_2) \in \mathcal{O}_1 \times \mathcal{O}_2 : g_1, g_2 \in C_G(D) \text{ and } g_1 g_2 = g\}.$$

Obviously $C_g \subseteq B_g$. We will show that, for $g \in C_G(D)$, the integer $|B_g \setminus C_g|$ is divisible by p and hence is equal to zero in k. Since D centralizes g we may view B_g as a D-set via the action

$$D \times B_g \longrightarrow B_g$$
$$(h, (g_1, g_2)) \longmapsto (h g_1 h^{-1}, h g_2 h^{-1}).$$

The subset C_g consists of exactly all the fixed points of this action. The complement $B_g \setminus C_g$ therefore is the union of all D-orbits with more than one element. But since D is a p-group any D-orbit has a power of p many elements. □

Literally the same reasoning works for any subgroup $C_G(D) \subseteq H \subseteq N_G(D)$ and shows that we may view s also as a homomorphism of k-algebras

$$s_{D,H} := s: \quad Z(k[G]) \longrightarrow Z(k[H]).$$

It is called the *Brauer homomorphism* (of G with respect to D and H). For any $e \in E(G)$ its image $s(e)$ either is equal to zero or is an idempotent. We want to establish a criterion for the nonvanishing of $s(e)$.

Lemma 5.5.2 *For any $\mathcal{O} \in \mathcal{O}(G)$ we have:*

i. *$\mathcal{O} \cap C_G(D) \neq \emptyset$ if and only if D is contained in a defect group of $\mathcal{O}$;*
ii. *if D is a defect group of $\mathcal{O}$, then $\mathcal{O} \cap C_G(D)$ is a single conjugacy class in $N_G(D)$;*
iii. *let $o \in \mathcal{O}(N_G(D))$ such that $o \subseteq \mathcal{O}$; if D is a defect group of o then D also is a defect group of $\mathcal{O}$ and $o = \mathcal{O} \cap C_G(D)$.*

Proof i. First let $x \in \mathcal{O} \cap C_G(D)$. Then $D \subseteq C_G(X)$, and D is contained in a p-Sylow subgroup of $C_G(x)$. Conversely, let D be contained in a p-Sylow subgroup of $C_G(x)$ for some $x \in \mathcal{O}$. Then D centralizes x and hence $x \in \mathcal{O} \cap C_G(D)$.

ii. We assume that D is a p-Sylow subgroup of $C_G(x)$ for some $x \in \mathcal{O}$. Then $x \in \mathcal{O} \cap C_G(D)$. Let $y \in \mathcal{O} \cap C_G(D)$ be any other point and let $g \in G$ such that $x = gyg^{-1}$. Since D centralizes y the conjugate group gDg^{-1} centralizes x. It follows that D as well as gDg^{-1} are p-Sylow subgroups of $C_G(x)$. Hence we find an $h \in C_G(x)$ such that $hDh^{-1} = gDg^{-1}$. We see that $h^{-1}g \in N_G(D)$ and $(h^{-1}g)y(h^{-1}g)^{-1} = h^{-1}gyg^{-1}h = h^{-1}xh = x$.

iii. We now assume that D is a p-Sylow subgroup of $C_{N_G(D)}(x)$ for some $x \in o$. Then D centralizes x and hence $x \in \mathcal{O} \cap C_G(D)$. It follows from i. that D is contained in a defect group of $\mathcal{O}$ and, more precisely, in a p-Sylow subgroup P of $C_G(x)$. If we show that $D = P$ then D is a defect group of $\mathcal{O}$ and ii. implies that $o = \mathcal{O} \cap C_G(D)$. Let us therefore assume that $D \subsetneqq P$. Then, as in any p-group, also $D \subsetneqq N_P(D)$. We pick an element $h \in N_P(D) \setminus D$ and let Q denote the p-subgroup generated by D and h. We then have

$$D \subsetneqq Q \subseteq N_P(D) \subseteq P \subseteq C_G(x),$$

from which we deduce that $D \subsetneqq Q \subseteq C_{N_G}(x)$. But D was a p-Sylow subgroup of $C_{N_G}(x)$. This is a contradiction. □

Lemma 5.5.3 *Let $e \in E(G)$ with corresponding central character χ_e; if D is minimal with respect to $e \in I_D$ then we have:*

i. *$s(e) \neq 0$;*
ii. *$\chi_e(I_Q) = \{0\}$ for any proper subgroup $Q \subsetneqq D$;*
iii. *if $\chi_e(\hat{\mathcal{O}}) \neq 0$ for some $\mathcal{O} \in \mathcal{O}(G)$ then D is contained in a defect group of $\mathcal{O}$.*

Proof i. Let $S \subseteq \mathcal{O}(G)$ denote the subset of all conjugacy classes $\mathcal{O}$ such that D contains a defect group $D_{\mathcal{O}}$ of $\mathcal{O}$. By assumption we have

$$e = \sum_{\mathcal{O} \in S} a_{\mathcal{O}} \hat{\mathcal{O}} \quad \text{with } a_{\mathcal{O}} \in k.$$

Suppose that

$$s(e) = \sum_{\mathcal{O} \in S} a_{\mathcal{O}} s(\hat{\mathcal{O}}) = \sum_{\mathcal{O} \in S} a_{\mathcal{O}} \Bigg(\sum_{x \in \mathcal{O} \cap C_G(D)} x \Bigg) = 0.$$

It follows that, for any $\mathcal{O} \in S$, we have $a_{\mathcal{O}} = 0$ or $\mathcal{O} \cap C_G(D) = \emptyset$. The latter, by Lemma 5.5.2.i, means that D is not contained in a defect group of $\mathcal{O}$. We therefore obtain that

$$e = \sum_{\mathcal{O} \in S, D_{\mathcal{O}} \subsetneqq D} a_{\mathcal{O}} \hat{\mathcal{O}} \in \sum_{\mathcal{O} \in S, D_{\mathcal{O}} \subsetneqq D} I_{D_{\mathcal{O}}}.$$

Remark 5.3.9 then implies that $e \in I_{D_{\mathcal{O}}}$ for some $\mathcal{O} \in S$ such that $D_{\mathcal{O}} \subsetneqq D$. This contradicts the minimality of D.

ii. By Remark 5.3.9 it follows from $e \in I_D$ that $I_D e = Z_e$ is a local ring. Then $I_Q e$, for any subgroup $Q \subseteq D$, is an ideal in this local ring. But if $Q \neq D$ then the minimality property of D says that $e \notin I_Q e \subseteq I_Q$. It follows that $I_Q e \subseteq \mathrm{Jac}(Z_e)$ and hence that $\chi_e(I_Q) = \chi_e(I_Q e) = \{0\}$.

iii. Let $D_{\mathcal{O}} \subseteq G$ be a defect group of $\mathcal{O}$. Using Proposition 5.3.7 we have

$$e I_{D_{\mathcal{O}}} \subseteq I_D I_{D_{\mathcal{O}}} \subseteq \sum_{g \in G} I_{D \cap g D_{\mathcal{O}} g^{-1}}.$$

Because of $\hat{\mathcal{O}} \in I_{D_{\mathcal{O}}}$ the central character χ_e cannot vanish on the right-hand sum. Hence, by ii., there must exist a $g \in G$ such that $D \subseteq g D_{\mathcal{O}} g^{-1}$. □

Lemma 5.5.4 *Let $e \in E(G)$ and $\mathcal{O} \in \mathcal{O}(G)$; if the p-subgroup P is normal in G then we have*:

i. *If $e \in I_P$ then $\delta(P)$ is a vertex of the $k[G \times G]$-module $k[G]e$*;
ii. *if $e \in I_Q$ for some p-subgroup $Q \subseteq G$ then $P \subseteq Q$*;
iii. *if $\mathcal{O} \cap C_G(P) = \emptyset$ then $\hat{\mathcal{O}} \in \mathrm{Jac}(Z(k[G]))$*.

Proof i. According to the proof of Corollary 5.3.12 the indecomposable $k[G \times G]$-module $k[G]e$ is relatively $k[P \times P]$-projective. Lemma 4.3.1 then implies that $k[G]e$ as a $k[G \times G]$-module and some indecomposable direct summand of $k[G]e$ as a $k[P \times P]$-module have a common vertex. But by Lemma 5.4.3.ii the summands in the decomposition

$$k[G] = \bigoplus_{y \in P \backslash G / P} k[PyP]$$

are indecomposable $k[P \times P]$-modules having the vertex

$$(1, y)^{-1} \delta\big(P \cap y P y^{-1}\big)(1, y).$$

It follows that $k[G]e$ has a vertex of the form $\delta(P \cap y P y^{-1})$ for some $y \in G$. Since P is assumed to be normal in G the latter group is equal to $\delta(P)$.

ii. We know from Proposition 5.3.11 that any $k[G]$-module M belonging to the e-block has a vertex contained in Q. Let X be a simple $k[G]$-module belonging to the e-block (cf. Proposition 1.7.4.i and Corollary 1.7.5 for the existence). By our assumption that P is a normal p-subgroup of G we may argue similarly as in the first half of the proof of Proposition 5.4.7 to see that P is contained in any vertex of X.

iii. Let χ_f, for $f \in E(G)$, be any central character, and let the p-subgroup $D_f \subseteq G$ be minimal with respect to $f \in I_{D_f}$. From ii. we know that $P \subseteq D_f$. On the other hand P cannot be contained in a defect group of $\mathcal{O}$ by Lemma 5.5.2.i. Hence D_f is not contained in a defect group of $\mathcal{O}$, which implies that $\chi_f(\hat{\mathcal{O}}) = 0$ by Lemma 5.5.3.iii. We see that $\hat{\mathcal{O}}$ lies in the kernel of every central character, and we deduce that $\hat{\mathcal{O}} \in \mathrm{Jac}(Z(k[G]))$. □

In the following we need to consider the ideals I_P for the same p-subgroup P in various rings $Z(k[H])$ for $P \subseteq H \subseteq G$. We therefore write $I_P(k[H])$ in order to refer to the ideal I_P in the ring $Z(k[H])$.

We fix a subgroup

$$DC_G(D) \subseteq H \subseteq N_G(D)$$

and consider the corresponding Brauer homomorphism

$$s_{D,H}\colon \quad Z\big(k[G]\big) \longrightarrow Z\big(k[H]\big).$$

Let $e \in E(G)$ be such that $s_{D,H}(e) \neq 0$ (for example, by Lemma 5.5.3.i, if D is minimal with respect to $e \in I_D(k[G])$). We may decompose

$$s_{D,H}(e) = e_1 + \cdots + e_r$$

uniquely into a sum of primitive idempotents $e_i \in E(H)$ (cf. Proposition 1.5.5).

Lemma 5.5.5 *If $e \in I_D(k[G])$ then $e_1, \ldots, e_r \in I_D(k[H])$.*

Proof Again let $S \subseteq \mathcal{O}(G)$ denote the subset of all conjugacy classes $\mathcal{O}$ such that D contains a defect group of $\mathcal{O}$. By assumption e is a linear combination of the $\hat{\mathcal{O}}$ for $\mathcal{O} \in S$. Hence $s_{D,H}(e)$ is a linear combination of $\hat{o}$ for $o \in \mathcal{O}(H)$ such that $o \subseteq \mathcal{O}$ for some $\mathcal{O} \in S$. Any defect group of such an o is contained in a defect group of $\mathcal{O}$ and therefore in gDg^{-1} for some $g \in G$ (cf. Exercise 5.3.5.i). It follows that

$$e_1 + \cdots + e_r = s_{D,H}(e) \in \sum_{g \in G} I_{gDg^{-1} \cap H}\big(k[H]\big)$$

and consequently, by Corollary 5.3.8, that

$$e_i = e_i(e_1 + \cdots + e_r) \in \sum_{g \in G} I_{gDg^{-1} \cap H}\big(k[H]\big) \quad \text{for any } 1 \leq i \leq r.$$

Remark 5.3.9 now implies that for any $1 \leq i \leq r$ there is a $g_i \in G$ such that

$$e_i \in I_{g_i D g_i^{-1} \cap H}\big(k[H]\big).$$

Since D is normal in H we may apply Lemma 5.5.4.ii and obtain that $D \subseteq g_i D g_i^{-1} \cap H$ and hence, in fact, that $D = g_i D g_i^{-1} \cap H$. □

We observe that

$$k[H]s_{D,H}(e) = \bigoplus_{f \in E(H)} k[H]f(e_1 + \cdots + e_r) = \bigoplus_{i=1}^{r} k[H]e_i .$$

Since D is normal in H it follows from Lemmas 5.5.5 and 5.5.4.i that $\delta(D)$ is a vertex of the indecomposable $k[H \times H]$-module $k[H]e_i$ for any $1 \le i \le r$.

On the other hand, G as an $H \times H$-set decomposes into $G = H \cup (G \setminus H)$. Correspondingly, $k[G]$ as a $k[H \times H]$-module decomposes into $k[G] = k[H] \oplus k[G \setminus H]$. We let $\pi_H : k[G] \longrightarrow k[H]$ denote the associated projection map, which is a $k[H \times H]$-module homomorphism.

Lemma 5.5.6 $\pi_H(e) \in s_{D,H}(e) + \mathrm{Jac}(Z(k[H]))$.

Proof Let $e = \sum_{\mathcal{O} \in \mathcal{O}(G)} a_{\mathcal{O}} \hat{\mathcal{O}}$ with $a_{\mathcal{O}} \in k$. Then

$$\pi_H(e) = \sum_{\mathcal{O} \in \mathcal{O}(G)} a_{\mathcal{O}} \Bigl(\sum_{x \in \mathcal{O} \cap H} x \Bigr)$$

whereas $s_{D,H}(e) = \sum_{\mathcal{O} \in \mathcal{O}(G)} a_{\mathcal{O}} (\sum_{x \in \mathcal{O} \cap C_G(D)} x)$. It follows that $\pi_H(e) - s_{D,H}(e)$ is of the form

$$\sum_{o \in \mathcal{O}(H), o \cap C_G(D) = \emptyset} b_o \hat{o} \quad \text{with } b_o \in k.$$

Lemma 5.5.4.iii (applied to $D \subseteq H$) shows that this sum lies in $\mathrm{Jac}(Z(k[H]))$. □

Proposition 5.5.7 *The $k[H \times H]$-module $k[H]s_{D,H}(e)$ is isomorphic to a direct summand of $k[G]e$.*

Proof We have the $k[H \times H]$-module homomorphism

$$\begin{aligned} \alpha : \quad k[H]s_{D,H}(e) &\longrightarrow k[G]e \otimes_{k[H]} k[H]s_{D,H}(e) \\ v &\longmapsto e \otimes v. \end{aligned}$$

By Lemma 5.5.6 the composite $\beta := (\pi_H \otimes \mathrm{id}) \circ \alpha \in \mathrm{End}_{k[H \times H]}(k[H]s_{D,H}(e))$ satisfies

$$\beta(v) = (\pi_H \otimes \mathrm{id}) \circ \alpha(v) = \pi_H(e)v = \bigl(s_{D,H}(e) + z\bigr)v$$

for any $v \in k[H]s_{D,H}(e)$ and some element $z \in \mathrm{Jac}(Z(k[H]))$. We note that $(e_i + z)e_i = e_i + ze_i$, for any $1 \le i \le r$, is invertible in the local ring $Z(k[H])e_i$. Therefore the element

$$y := \left(\prod_{i=1}^{r}(e_i + ze_i)^{-1}\right) \times \left(\prod_{f\in E(H),\, f\neq e_i} 0\right) \in Z\big(k[H]\big)^{\times} s_{D,H}(e)$$

is well defined and satisfies

$$\big(s_{D,H}(e) + z\big)y = s_{D,H}(e).$$

Then

$$\big(s_{D,H}(e) + z\big)k[H]s_{D,H}(e) = \big(s_{D,H}(e) + z\big)yk[H]s_{D,H}(e) = k[H]s_{D,H}(e)$$

which implies that the map β is surjective. Since $k[H]s_{D,H}(e)$ is finite-dimensional over k the map β must be bijective. It follows that $k[H]s_{D,H}(e)$ is isomorphic to a direct summand of $k[G]e \otimes_{k[H]} k[H]s_{D,H}(e)$. Furthermore the latter is a direct summand of

$$\begin{aligned} k[G]e &= k[G]e \otimes_{k[H]} k[H] \\ &= \big(k[G]e \otimes_{k[H]} k[H]s_{D,H}(e)\big) \oplus \big(k[G]e \otimes_{k[H]} k[H]\big(1 - s_{D,H}(e)\big)\big). \quad \square \end{aligned}$$

Theorem 5.5.8 *For any $e \in E(G)$ and any defect group D_0 of the e-block we have:*

i. $e \in I_{D_0}(k[G])$;
ii. *if $H = N_G(D_0)$ then $s_{D_0,H}(e) \in E(H)$ and $B(e) = s_{D_0,H}(e)$.*

Proof With the notations from before Lemma 5.5.5, where $D \subseteq G$ is a p-subgroup which is minimal with respect to $e \in I_D(k[G])$, we know from Lemma 5.5.5 that $e_1, \dots, e_r \in I_D(k[H])$. Applying Lemma 5.5.4.i to the normal p-subgroup D of H we obtain that $\delta(D)$ is a vertex of the indecomposable $k[H \times H]$-modules $k[H]e_i$. But by Proposition 5.5.7 these $k[H]e_i$ are isomorphic to direct summands of $k[G]e$. Hence it follows from Lemma 4.3.1 that $D \subseteq gD_0g^{-1}$ for some $g \in G$. We see that $e \in I_D \subseteq I_{gD_0g^{-1}} = I_{D_0}$ which proves i.

In fact Corollary 5.3.12 says that D has to contain some defect subgroup of the e-block. We conclude that $D = gD_0g^{-1}$ and therefore that D_0 is minimal with respect to $e \in I_{D_0}(k[G])$ as well. This means that everywhere in the above reasoning we may replace D by D_0 (so that, in particular, $D_0C_G(D_0) \subseteq H \subseteq N_G(D_0)$). We obtain that D_0 also is a defect group of the e_i-blocks for $1 \leq i \leq r$. Let $H = N_G(D_0)$ so that $N_{G\times G}(\delta(D_0)) \subseteq H \times H$. By the Green correspondence in Theorem 4.3.6 the indecomposable $k[G \times G]$-module $k[G]e$ with vertex D_0 has, up to isomorphism, a unique direct summand, as a $k[H \times H]$-module, with vertex D_0. It follows that $r = 1$, which means that $s_{D_0,H}(e) = e_1 \in E(H)$, and that this summand is isomorphic to $k[H]e_1$, which means that $B(e) = e_1$. This proves ii. $\square$

Proposition 5.5.9 *For any $e \in E(G)$ and any defect group D_0 of the e-block we have:*

i. *If the p-Sylow subgroup* $Q \subseteq G$ *contains* D_0 *then there exists a* $g \in G$ *such that* $D_0 = Q \cap gQg^{-1}$;
ii. *any normal p-subgroup* $P \subseteq G$ *is contained in* D_0.

Proof i. By assumption and Exercise 4.2.2.iii the indecomposable $k[G \times G]$-module $k[G]e$ is relatively $k[Q \times Q]$-projective. Lemma 4.3.2 therefore implies that $k[G]e$ as a $k[Q \times Q]$-module has an indecomposable direct summand X with vertex $\delta(D_0)$. Applying Lemma 5.4.3.ii to the decomposition

$$k[G] = \bigoplus_{y \in Q\backslash G/Q} k[QyQ]$$

we see that X also has a vertex of the form $(1, y)^{-1}\delta(Q \cap yQy^{-1})(1, y)$ for some $y \in G$. It follows that

$$\delta(D_0) = (q_1, q_2)(1, y)^{-1}\delta\bigl(Q \cap yQy^{-1}\bigr)(1, y)(q_1, q_2)^{-1}$$

for some $(q_1, q_2) \in Q \times Q$. We obtain on the one hand that $|D_0| = |Q \cap yQy^{-1}|$ and on the other hand that

$$q_1^{-1}dq_1 = yq_2^{-1}d\bigl(yq_2^{-1}\bigr)^{-1} \in Q \cap yQy^{-1} \quad \text{for any } d \in D_0.$$

The element $g := q_1yq_2^{-1}$ then lies in $C_G(D_0)$. Hence $D_0 \subseteq Q \cap gQg^{-1}$. Moreover

$$\begin{aligned} |Q \cap gQg^{-1}| &= |Q \cap q_1yq_2^{-1}Qq_2y^{-1}q_1^{-1}| \\ &= |q_1^{-1}Qq_1 \cap yq_2^{-1}Qq_2y^{-1}| = |Q \cap yQy^{-1}| \\ &= |D_0|. \end{aligned}$$

ii. Since P is contained in any p-Sylow subgroup of G it follows from i. that $P \subseteq D_0$. □

Theorem 5.5.10 *Let* $D \subseteq G$ *be a p-subgroup and fix a subgroup* $DC_G(D) \subseteq H \subseteq N_G(D)$. *We then have:*

i. $E_{\geq D}(H) = E(H)$;
ii. *the Brauer correspondence* $\beta : E(H) \longrightarrow E_{\geq D}(G)$ *is characterized in terms of central characters by*

$$\chi_{\beta(f)} = \chi_f \circ s_{D,H} \quad \text{for any } f \in E(H);$$

iii. *if* $e \in E(G)$ *such that* $s_{D,H}(e) \neq 0$ *then* $e \in \mathrm{im}(\beta)$ *and*

$$s_{D,H}(e) = \sum_{f \in \beta^{-1}(e)} f;$$

iv. $E_D(G) \subseteq \mathrm{im}(\beta)$.

Proof i. Since D is normal in H this is immediate from Proposition 5.5.9.ii.

ii. For any $f \in E(H)$ the composed k-algebra homomorphism $\chi_f \circ s_{D,H}$ must be a central character of $k[G]$, i.e. $\chi_f \circ s_{D,H} = \chi_e$ for some $e \in E(G)$. Since $1 = \chi_e(e) = \chi_f(s_{D,H}(e))$ we have $s_{D,H}(e) \neq 0$. Let $s_{D,H}(e) = e_1 + \cdots + e_r$ with $e_i \in E(H)$. Then $1 = \chi_f(e_1) + \cdots + \chi_f(e_r)$ and hence $f = e_j$ for some $1 \leq j \leq r$. By Proposition 5.5.7 the indecomposable $k[H \times H]$-module $k[H]f$ is isomorphic to a direct summand of $k[G]e$. It follows that $e = \beta(f)$.

iii. Let $s_{D,H}(e) = e_1 + \cdots + e_r$ with $e_i \in E(H)$. Again it follows from Proposition 5.5.7 that $e = \beta(e_1) = \cdots = \beta(e_r)$. On the other hand, if $e = \beta(f)$ for some $f \in E(H)$ then we have seen in the proof of ii. that necessarily $f = e_j$ for some $1 \leq j \leq r$.

iv. If $e \in E_D(G)$ then D is minimal with respect to $e \in I_D(k[G])$ by Corollary 5.3.12 and Theorem 5.5.8.i. Hence $s_{D,H}(e) \neq 0$ by Lemma 5.5.3.i. □

Proposition 5.5.11 *For any $e \in E(G)$ and any defect group D_0 of the e-block we have*:

i. *If $\chi_e(\hat{\mathcal{O}}) \neq 0$ for some $\mathcal{O} \in \mathcal{O}(G)$ then D_0 is contained in a defect group of $\mathcal{O}$*;
ii. *there is an $\mathcal{O} \in \mathcal{O}(G)$ such that $\chi_e(\hat{\mathcal{O}}) \neq 0$ and D_0 is a defect group of $\mathcal{O}$.*

Proof i. By Corollary 5.3.12 and Theorem 5.5.8.i the defect group D_0 is minimal with respect to $e \in I_{D_0}(k[G])$. The assertion therefore follows from Lemma 5.5.3.iii.

ii. Since $e \in I_{D_0}(k[G])$ we have

$$e = \sum_{\mathcal{O} \in S} a_{\mathcal{O}} \hat{\mathcal{O}} \quad \text{with } a_{\mathcal{O}} \in k$$

where $S \subseteq \mathcal{O}(G)$ is the subset of all $\mathcal{O}$ such that D_0 contains a defect group of $\mathcal{O}$. Then $1 = \chi_e(e) = \sum_{\mathcal{O} \in S} a_{\mathcal{O}} \chi_e(\hat{\mathcal{O}})$. Hence there must exist an $\mathcal{O} \in S$ such that $\chi_e(\hat{\mathcal{O}}) \neq 0$. This $\mathcal{O}$ has a defect group contained in D_0 but by i. also one which contains D_0. It follows that D_0 is a defect group of $\mathcal{O}$. □

References

1. Alperin, L.: Local Representation Theory. Cambridge University Press, Cambridge (1986)
2. Craven, D.: The modular representation theory of finite groups. Univ. Birmingham, Thesis (2006)
3. Curtis, C., Reiner, I.: Methods of Representation Theory, Vol. I. Wiley, New York (1981)
4. Dixon, J., Puttaswamaiah, B.: Modular Representations of Finite Groups. Academic Press, New York (1977)
5. Dornhoff, L.: Group Representation Theory, Part B: Modular Representation Theory. Dekker, New York (1972)
6. Feit, W.: The Representation Theory of Finite Groups. North-Holland, Amsterdam (1982)
7. Isaacs, I.M.: Character Theory of Finite Groups. Academic Press, New York (1976)
8. Lam, T.Y.: A First Course in Noncommutative Rings. Springer, Heidelberg (1991)
9. Schneider, P.: Die Theorie des Anstiegs. Course at Münster, 2006/2007. Available at www.math.uni-muenster.de/u/schneider/publ/lectnotes/
10. Serre, J.P.: Linear Representations of Finite Groups. Springer, Heidelberg (1977)

P. Schneider, *Modular Representation Theory of Finite Groups*,
DOI 10.1007/978-1-4471-4832-6,

Index

P. Schneider, *Modular Representation Theory of Finite Groups*,
DOI 10.1007/978-1-4471-4832-6, © Springer-Verlag London 2013